THE PROTEIN MOLECULE

Cover Figure:
Schematic structure of goose, hen and T4 phage lysozymes
(Weaver, L.H. *et al.* (1985) *J. Mol. Evol., 21,* 97).

THE PROTEIN MOLECULE

MOLECULE

Conformation, Stability
and Folding

KOZO HAMAGUCHI

Emeritus Professor,
Osaka University,
Osaka, Japan

With 87 Figures and 43 Tables

JAPAN SCIENTIFIC SOCIETIES PRESS
Tokyo

SPRINGER-VERLAG
Berlin Heidelberg New York London Paris
Tokyo Hong Kong Barcelona Budapest

Supported in part by the Ministry of Education, Science and Culture under Grant-in-Aid for Publication of Scientific Research Result.

Published jointly by:
JAPAN SCIENTIFIC SOCIETIES PRESS Tokyo
ISBN 4-7622-5706-0

and

SPRINGER-VERLAG Berlin Heidelberg New York London Paris
Tokyo Hong Kong Barcelona Budapest
ISBN 3-540-55915-9 SPRINGER-VERLAG Berlin Heidelberg New York
ISBN 0-387-55915-9 SPRINGER-VERLAG New York Berlin Heidelberg

Sole distribution rights outside Japan granted to SPRINGER-VERLAG

GENETICS INSTITUTE⁺

002767

INFORMATION CENTER

Printed in Japan

Preface

Physicochemical studies of proteins started as one of the fields of colloid science, since the dimensions of most proteins are within the range of the dimensions of colloidal particles. Most early studies of proteins were mainly concerned with determination of the size and shape of the protein molecule and with the properties of charged hydrophilic colloidal particles.

Since 1950, beginning with the determination of the amino acid sequence of insulin by F. Sanger and H. Tuppy (1951) and the proposal of the α-helix and β-structure by L. Pauling (1951), many reports on the primary and secondary structures of proteins have appeared. Optical rotation, optical rotatory dispersion, circular dichroism and infrared absorption have been used to examine the presence of α-helical structures in protein molecules. Synthetic polypeptides, such as poly-L-glutamic acid and poly-L-lysine, and synthetic co-polypeptides were used as reference substances to estimate the secondary structure of protein molecules. The classification of protein structure into primary, secondary and tertiary structures was proposed by K. Linderstrøm-Lang and J.A. Schellman in 1959.

In parallel with these studies of protein secondary structure, the factors responsible for stabilizing the protein molecule into a compact globular shape were also investigated extensively. For this purpose, many studies of protein denaturation by temperature, pH and various denaturing agents were carried out, even though it had been found as early as the 1930s that protein denaturation is reversible, at which time the calculation of thermodynamic parameters for the unfolding reactions had been attempted by M.L. Anson, A.E. Mirsky and M. Kunitz. Until the first half of the 1950s, electrostatic interactions and hydrogen bonds were considered to be the main factors stabilizing the protein molecule. This seemed a natural consequence, because proteins were initially treated as colloidal particles with surface charges, and the importance of hydrogen bonds had been emphasized since Pauling's original proposal of the α-helix and β-structure. However, hydrophobic residues comprise about 50% of the constituent amino acid risidues in most proteins. On this basis, W. Kauzmann considered that the interactions between these hydrophobic residues would be important for stabilizing the protein molecule, and proposed the term "hydrophobic bond" in 1954. In 1959 he further clarified the properties of these hydrophobic interactions, estimating their magnitude on the basis of changes in the thermodynamic parameters of transfer of various non-polar compounds from water to organic solvents, and published a review of the various factors that stabilize the protein molecule—hydrogen bonds, hydrophobic bonds, electrostatic interactions and conformational entropy.

In the late 1950s, C.B. Anfinsen made the important discovery that the ribonuclease A molecule denatured by 8 M urea and with all its disulphide bonds reduced, refolds into the original structure and regains its original activity by air oxidation after removal of the denaturant. This experiment demonstrated that the conformation of the native protein represents a state of minimum free energy, and that when the linear sequence of the polypeptide is synthesized *in vivo*, the native structure is formed spontaneously. This principle is referred to as "Anfinsen's dogma".

Mention should also be made of a very important study that employed the hydrogen-deuterium exchange method developed by K. Linderstrøm-Lang in the 1950s, which showed that the protein molecule always fluctuates.

The concept of specificity in reactions involving proteins, such as enzymatic and immunological reactions, was also established in the late 1950s. The study by F.M. Richards (1959) on interactions between the S-protein and S-peptide obtained by limited proteolysis of ribonuclease A with subtilisin was also important for understanding the specificity and complementarity of protein reactions.

Thus, ideas on how the protein molecule is constituted, how it is stabilized and how it is folded were well defined before the detailed structure of the protein molecule was elucidated.

Since 1960, beginning with the X-ray crystallographic analysis of myoglobin by J. Kendrew *et al.* (1960), many papers on the crystallographic structures of proteins have appeared. Needless to say, these results have had a profound effect on structural and biochemical studies, and much progress has also been made in studies of proteins in solution. However, the fundamental concepts of protein stability and folding have not necessarily changed, although more detailed information is now being advanced. The 1960s and 1970s can be called the years of protein structure elucidation.

The 1980s were the years new proteins were produced artificially. Site-specific modification by synthetic and recombinant DNA technology has been used actively to clarify the stability and function of proteins. So-called protein engineering (Drexler, 1981) is an interesting and powerful method for clarifying the role of a specific residue in protein stability and function. However, the results of these investigations have led to the conclusion that new proteins produced artificially are never superior to proteins that have naturally evolved over periods as long as four billion years. Even if one property of a protein is improved by replacement of a specific residue, impairment of another property may result.

The functions of proteins are closely related to their conformations. This is one of the most interesting but difficult subjects in studies of proteins. The present book concentrates mainly on the conformation, stability and folding of monomeric proteins, and describes the molecular basis of protein functions.

Chapter 1 deals with the physicochemical properties of the constituent amino acid residues of proteins. The hydrophilicity, hydrophobicity and ionization characteristics of the side chains of the twenty amino acid residues play a decisive role in determining the protein conformation. Chapters 2 and 3 describe the secondary and tertiary structures of proteins obtained on the basis of X-ray crystallographic analyses, and the general rules of protein conformation.

Various non-covalent interactions that stabilize the protein molecule are described in Chapter 4, and data on the stability of proteins obtained on the basis of denaturation studies in Chapter 5. Thus the subjects treated in these two chapters are closely related. Chapter 5 also provides details on the stability of proteins inferred from replacement of a specific residue by protein engineering.

Chapter 6 is concerned with fluctuations of the protein molecule. As described above, protein fluctuations were first investigated in the pioneering work of Linderstrøm-Lang, who measured the change with time in the density of a mixture of light and heavy water produced by exchange of the exchangeable protons of a protein with deuterium in the solvent. Nowadays, the exchange of a specific hydrogen atom with deuterium can be followed by the nuclear magnetic resonance (NMR) method and crystallographic analysis. Fluctuations in protein structure are now considered to be closely related to protein function.

Chapter 7 describes how the native structure of the protein molecule is formed from its linear structure *in vitro* and *in vivo*. This chapter should be of particular interest to those whose research deals with the various aspects of protein structure and function.

Finally, Chapter 8 deals with the molecular basis of the enzymatic function of lysozyme, one of the most thoroughly studied enzymes, including the ionization behaviour of the catalytic groups and those that participate in the binding of substrate, and the details of their interactions with substrate.

Acknowledgements

I wish to thank many investigators who assisted me, in particular: Dr. B.J. Sutton, King's College, University of London, for his critical comments and invaluable advice on the entire manuscript, and Dr. Y. Goto, Osaka University and Dr. Y. Kawata, Tottori University, for their comments on Chapters 4, 5 and 7. I also wish to thank Mrs. Y. Kawata for her considerable help in the preparation of the manuscript.

I am most grateful to those authors and publishers who allowed me to reproduce published figures and due acknowledgement is made in each case in the text.

The publication of this book has been supported by a grant-in-aid for publication of scientific results from the Ministry of Education, Science and Culture of Japan.

Contents

1

Physicochemical Properties of Amino Acid Side Chains

1-1. CONSTITUENT AMINO ACIDS

Proteins are constituted from twenty different amino acids, which are joined together by amide links, known as peptide bonds, and have the following structure.

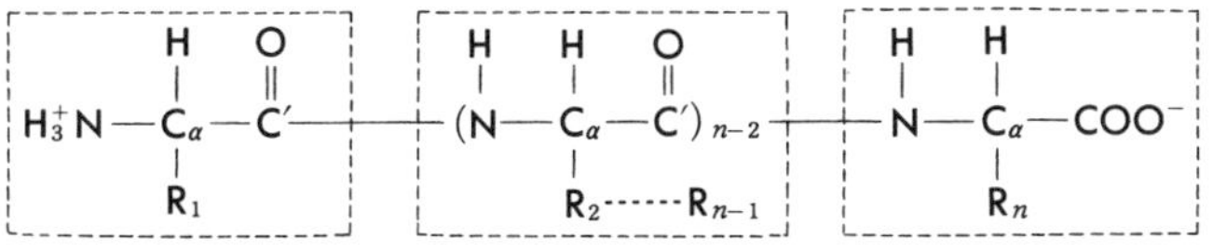

(Amino terminal) (Carboxy terminal)

Each unit surrounded by a dotted box is called the amino acid residue. R_1, R_2, ------R_n are the side chains. The carbonyl carbon of the peptide bond is often represented by C′ in order to distinguish it from the α-carbon. All protein molecules have the repeating sequence,

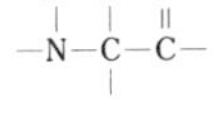

$$-\overset{\text{H}}{\underset{|}{N}}-\overset{\text{H}}{\underset{|}{C}}-\overset{\text{O}}{\overset{||}{C}}-$$

, in their backbone.

TABLE 1-1
Constituent Amino Acids

Amino acid	Three-letter symbol	One-letter symbol	Side chains
Hydrophobic amino acid residues			
Glycine	Gly	G	H
Alanine	Ala	A	CH_3
Valine	Val	V	CH_3 CH_3 CH
Leucine	Leu	L	CH_3 CH_2-CH CH_3
Isoleucine	Ile	I	CH_3 CH_2 CH CH
Methionine	Met	M	CH_3 CH_2-CH_2-S
Proline	Pro	P	CH_2 CH_2 CH_2 $N-CH$
Phenylalanine	Phe	F	$CH-CH$ CH_2-C CH $CH-CH$
Tryptophan	Trp	W	$CH-CH$ HC CH $C-C$ HN C CH CH_2

TABLE 1-1 (continued)

Hydrophilic amino acid residues

Serine	Ser	S	CH_2-OH
Threonine	Thr	T	CH with OH and CH_3
Asparagine	Asn	N	CH_2-CONH_2
Glutamine	Gln	Q	$CH_2-CH_2-CONH_2$

Ionizable amino acid residues

Aspartic acid	Asp	D	$CH_2-COO^{\ominus}$
Glutamic acid	Glu	E	$CH_2-CH_2-COO^{\ominus}$
Cysteine	Cys	C	CH_2-SH
Tyrosine	Tyr	Y	CH_2-C ring, $C-OH$ or $-O^{\ominus}$
Histidine	His	H	CH_2-C ring, or $NH^{\oplus}$
Lysine	Lys	K	$CH_2-CH_2-CH_2-CH_2-NH_3^{\oplus}$

4

TABLE 1-1 (continued)

Arginine	Arg	R	

$-CH_2-CH_2-CH_2-NH-C^{\oplus}\begin{smallmatrix}NH_2\\NH_2\end{smallmatrix}$

The sequence of the amino acid residues in proteins is called the primary structure. The sequence is written with the amino terminal end to the left and the carboxy terminal end to the right, and is numbered from the amino terminal end. For instance, Lys 21 represents a lysine residue, which is the twenty-first residue from the amino terminal end.

The twenty most commonly occurring amino acid residues which constitute protein molecules are given in Table 1-1. In this table, the amino acid residues are classified into hydrophobic, hydrophilic and ionizable groups. As described below, this classification is helpful for understanding the physicochemical properties of proteins. No simple classification is satisfactory, as many side chains have characteristics of more than one group.

In addition to the commonly occurring amino acid residues, other more unusual ones are found in some proteins. These particular amino acids are derivatives of the more common amino acids. For instance, 4-hydroxyproline (I) and 5-hydroxylysine (II) are found in collagen, N-methyllysine (III) in myosin, and γ-carboxyglutamic acid (IV) in prothrombin and some Ca-binding proteins. Elastin contains desmosine (V), a derivative of lysine.

Almost all the proteins found in nature are constituted from L-amino acids. Figure 1-1 shows L- and D-amino acids. For L-amino acids, the groups appear in the clockwise order CO, R, N (CO: COOH, R: side chain, N: NH_2) when viewed along the bond from H to C, whereas for D-amino acids they appear in counterclockwise order.

The physicochemical properties of the side chains of amino acid residues and the amino acid sequence are dominant factors in

$$HO-\overset{H}{\underset{\underset{\underset{\overset{|}{H}}{N}}{H_2C_5{}^2}}{C_4{}^{-}}}{}_{3}CH_2$$
$$H_2C_5{}^2CH\text{-}COOH$$

(I)

$$H_2N-\overset{6}{CH_2}-\overset{5}{\underset{\underset{OH}{|}}{CH}}-\overset{4}{CH_2}-\overset{3}{CH_2}-\overset{2}{\underset{\underset{NH_2}{|}}{CH}}-\overset{1}{COOH}$$

(II)

$$CH_3-NH-\overset{6}{CH_2}-\overset{5}{CH_2}-\overset{4}{CH_2}-\overset{3}{CH_2}-\overset{2}{\underset{\underset{NH_2}{|}}{CH}}-\overset{1}{COOH}$$

(III)

$$\overset{HOOC}{\underset{HOOC}{>}}CH_2-CH_2-\underset{\underset{NH_2}{|}}{CH}-COOH$$

(IV)

$$\overset{H_2N}{\underset{HOOC}{>}}CH-(CH_2)_2-$$

H_2N COOH
      \ /
       CH
       |
     (CH_2)_3
                    NH_2
$$CH-(CH_2)_2-\underset{}{}-(CH_2)_2-CH\overset{}{<}COOH$$
       N
       |
     (CH_2)_4
       |
       CH
   NH_2 COOH

(V)

determining the structure and function of proteins. The ionization characteristics, hydrophilicity and hydrophobicity of the side chains are described below.

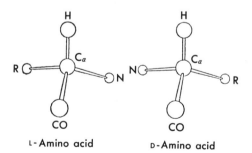

Fig. 1-1.   L-Amino acid and D-amino acid.

1-2. IONIZABLE AMINO ACID RESIDUES

The fundamental quantity which describes the ionization behaviour is the ionization constant (K_a) or pK_a. The standard free energy change of ionization ($\Delta G°$) is related to K_a by the following equation

$$\Delta G° = -RT \cdot \ln K_a = 2.303 \cdot RT \cdot pK_a. \tag{1.1}$$

The standard enthalpy change ($\Delta H°$), entropy change ($\Delta S°$) and heat capacity change ($\Delta C_p°$) of ionization are obtained by the following equations

$$\Delta H° = RT^2 \frac{\partial \ln K_a}{\partial T} = -R\frac{\partial \ln K_a}{\partial (1/T)} = 2.303\ R\frac{\partial pK_a}{\partial (1/T)} \tag{1.2}$$

$$\Delta S° = -\frac{\partial G°}{\partial T} = \frac{\Delta H° - \Delta G°}{T} \tag{1.3}$$

$$\Delta C_p° = \frac{\partial \Delta H°}{\partial T}. \tag{1.4}$$

The ionizable groups include the N-terminal α-amino group and C-terminal α-carboxyl group, in addition to the side chains of Asp, Glu, His, Lys, Cys, Tyr and Arg given in Table 1-1. The ionization constant is affected by various factors such as the electronic structure, the properties of the solvent, the microenvironment (particularly important in a folded protein structure) and steric effects.

The carboxyl group of Asp or Glu, $R-C{\overset{O}{\underset{OH}{}}}$, is far more easily ionizable than the corresponding alcohol $R-C{\overset{H_2}{\underset{OH}{}}}$. The pK_a value of an alcohol is more than 6 times larger than that of the corresponding carboxyl group. The ionized form of the carboxyl group, $R-C{\overset{O}{\underset{O}{}}}$, becomes more stable than its unionized form, $R-C{\overset{O}{\underset{OH}{}}}$, owing to its resonance structure. Thus the carboxyl is able to liberate a proton owing to the decrease in energy resulting from ionization. Such stabilization by formation of a resonance structure does not occur upon ionization of an alcohol.

The guanidinium ion has a symmetrical planar structure, the three C–N bonds are equivalent, and each N–C–N bond angle is 120°C. However, on release of a proton to form guanidine, this symmetrical resonance structure is lost. Thus the side chain guanidinium ion of Arg is very weakly acidic owing to stabilization of the ionic structure by resonance.

Guanidinium ion   Guanidine

Likewise, the imidazolium ion of His has a more symmetrical resonance structure than its conjugate base, imidazole.

Imidazolium ion   Imidazole

On the basis of the results obtained by analysis of titration data for ribonuclease A (Nozaki & Tanford, 1967c) and for hen egg-white lysozyme (Roxby & Tanford, 1971) in 6 M guanidine hydrochloride, the pK_a value of each ionizable group has been determined. As described in Chapter 5, proteins assume random coil conformations, and the distance between charged groups becomes large in 6 M guanidine hydrochloride. Furthermore, the electrostatic interactions between charges are assumed to be small in a medium of such high ionic strength. The pK_a values thus determined for ribonuclease A and lysozyme in 6 M guanidine hydrochloride were found to be in good agreement with each other and with those for small model compounds (Nozaki & Tanford, 1967c). These pK_a values are given in Table 1-2 together with the enthalpy changes of ionization. These values represent the pK_a values of ionizable groups in water at 25°C when electrostatic interactions between charges are absent, and are very important for understanding the fundamental ionization behaviour of ionizable groups in proteins. These values are called the "normal" pK_a values.

As can be seen in Table 1-2, the side chains of Asp and Glu, and the C-terminal α-carboxyl group each have a negative charge, and the side chains of Lys and Arg, and the N-terminal α-amino group each have a positive charge at physiological pH values. Only the imidazole side chain of His is titratable at neutral pH values. This is one of the reasons why most enzymes which function at neutral pH utilize His as one of the catalytic groups.

As also shown in the table, the enthalpy changes for the ionization of carboxyl groups are close to zero. Therefore, the free energy change of ionization for carboxyl groups is governed by the negative entropy change of ionization. For the ionization of acetic acid

$$CH_3COOH + H_2O \longrightarrow CH_3COO^- + H_3O^+,$$

the enthalpy change is close to zero and the entropy change is -30 cal/deg/mol (on the basis of unitary unit (see Section 1-3)) and the heat capacity change is -34 cal/deg/mol. When the carboxyl group is ionized, two charges, $-COO^-$ and H_3O^+, are newly produced and the water molecules around these charges are compressed and oriented (this phenomenon is called "electrostriction"); thus the overall entropy is decreased on ionization.

For the ionization of a basic group such as an amino group

$$RNH_3^+ + H_2O \longrightarrow RNH_2 + H_3O^+$$

the number of charges does not change. Thus the entropy change for ionization of an amino group is smaller than that for a carboxyl group. The effect on the water structure is larger for the smaller H_3O^+ ion than for the larger NH_3^+ ion, and thus the entropy change on ionization of the amino group is small and negative. Electrostriction causes a decrease in volume by 10 ml/mol for ionization of the carboxyl group and by 4 ml/mol for that of the amino group.

Measurements of the titration curves at two or more temperatures will allow estimation of the enthalpy change of ionization using Eq. (1.2). When titration of a protein is started from a very low pH value, the carboxyl group is titrated first, followed by the ionization of the imidazole group. Although the enthalpy change for the ionization is close to zero as described above, that for the imidazole group is $\simeq 7$ kcal/mol. Therefore, measurement of the titration curves of a protein at different temperatures is useful for determining the pH at which ionization of the imidazole group begins.

The ionization behaviour of ionizable groups in proteins depends strongly on the conditions of the environment in which they are located, and the pK_a values sometimes differ greatly from the values listed in Table 1-2. For instance, the pK_a value of an ionizable group depends on the dielectric constant and ionic strength of the medium. The pK_a values for acetic acid, tris-(hydroxymethyl)aminomethane, benzoyl arginine and glycine in water-dioxane mixtures are given in Table 1-3. As the dielectric constant decreases, the pK_a value of the carboxyl group increases, but the pK_a values of the cations are rather insensitive.

The work of charging a sphere to give a spherical ion with the charge Ze, where e is the proton charge and Z is the valence of the

TABLE 1-2

The Normal Ionization Constants and Heat of Ionization of Ionizable Groups in Proteins (25°C)

Ionizable groups		pK_a	$\Delta H°$ (kcal/mol)
C-terminus	α-COOH	3.4–3.8	$-1 - +1$
Asp	β-COOH	3.9–4.0	$-1 - +1$
Glu	γ-COOH	4.4–4.5	$-1 - +1$
His	HN$\overset{\oplus}{\cdots}$NH	6.3–6.6	$-6 - +7$
N-terminus	α-NH$_3^+$	7.4–7.5	$+9 - +11$
Lys	ε-NH$_3^+$	10.0–10.4	$+10 - +11$
Cys	-SH	(7.5–9.5)	$+6 - +7$
Tyr	-OH	9.6–10.0	$+6 - +7$
Arg	$-NHC\overset{NH_2}{\underset{NH_2}{\oplus}}$	> 12.5	$+12 - +13$

TABLE 1-3

pK$_a$ Values for Acetic Acid, Tris, Benzoylarginine and Glycine in Water-dioxane Mixtures (25°C)

Weight % dioxane	pK$_a$				
	Acetic acid	Tris[a]	Benzoyl-arginine	Glycine	
				-COOH	-NH$_3^+$
0	4.76	8.0	3.34	2.35	9.78
20	5.29	8.0	—	2.63	9.29
45	6.31	8.0	—	3.11	8.49
50	—	8.0	4.59	—	—
70	8.34	8.0	4.60	3.96	7.42

[a]Tris(hydroxymethyl)aminomethane.

Fersht, A. (1985) "Enzyme Structure and Mechanism," 2nd Ed., p.173, W.H. Freeman & Co., New York.

ion, *i.e.*, the electrical free energy, is given by the following equation

$$G° = \frac{NZ^2e^2}{2Db},\qquad(1.5)$$

where b is the radius of the ion, D is the dielectric constant of the medium, and N is Avogadro's number.

Since two ions are produced on ionization of the carboxyl group, the electrical free-energy change on ionization in water is expressed by the equation

$$(\Delta G°)_{H_2O} = \frac{Ne^2}{2D_0}\left(\frac{1}{b_1}+\frac{1}{b_2}\right),\qquad(1.6)$$

where D_0 is the dielectric constant of water, and b_1 and b_2 are the radii of the two ions produced. For ionization of the carboxyl group in another medium (A) of dielectric constant D, the electrical free-energy change is expressed by the equation

$$(\Delta G°)_A = \frac{Ne^2}{2D}\left(\frac{1}{b_1}+\frac{1}{b_2}\right).\qquad(1.7)$$

The difference in the electrical free-energy change between ionization of the carboxyl group in water and that in medium A becomes

$$(\Delta G^\circ)_A - (\Delta G^\circ)_{H_2O} = \frac{Ne^2}{2} \left(\frac{1}{D} - \frac{1}{D_0} \right) \left(\frac{1}{b_1} + \frac{1}{b_2} \right). \tag{1.8}$$

If D_0 and D are taken as 80 and 18 (corresponding to the dielectric constant of 70% dioxane), respectively, and b_1 and b_2 are taken as 2 Å, we have

$$(\Delta G^\circ)_A - (\Delta G^\circ)_{H_2O} \simeq 3 \times 10^{11} \text{ erg/mol}$$
$$\simeq 7.2 \text{ kcal/mol.}$$

This gives a pK_a value of 5.3 using Eq. (1.1). As shown in Table 1-3, the difference between the pK_a value of acetic acid in water and that in dioxane is 3.58. Although the above calculation is only approximate, it explains roughly the change in the pK_a value of the carboxyl group with a change in the dielectric constant of the medium.

The electrical free-energy change for ionization of the amino group in water is given by the following equation

$$(\Delta G^\circ)_{H_2O} = \frac{Ne^2}{2D_0} \left(\frac{1}{b_1} - \frac{1}{b_2} \right), \tag{1.9}$$

where b_1 is the radius of a proton and b_2 is the radius of the NH_3^+ ion. The difference between the electrical free-energy change in water and that in another medium A is given by

$$(\Delta G^\circ)_A - (\Delta G^\circ)_{H_2O} = \frac{Ne^2}{2D_0} \left(\frac{1}{D} - \frac{1}{D_0} \right) \left(\frac{1}{b_1} - \frac{1}{b_2} \right). \tag{1.10}$$

Since b_2 is larger than b_1, it is expected that the pK_a value for the amino group would be larger in a medium of lower dielectric constant than in water. In fact, however, the pK_a values for the cationic groups are less sensitive to a change in the dielectric constant in comparison with the pK_a value of the carboxyl group (Table 1-3), and the above treatment does not completely explain the change in the pK_a value with the change in the dielectric constant.

The pK_a value also depends on the ionic strength of the medium. The change in pK_a in the presence of salts can be treated similarly to the above situation. The electrical free energy in charging a sphere of radius b at ionic strength I is given by

$$G° = \frac{NZ^2e^2}{2D} \left(\frac{1}{b} - \frac{\varkappa}{1+\varkappa a} \right)$$

$$\varkappa = \sqrt{\frac{4\pi e^2}{DkT} \frac{2N}{1,000} I}, \qquad (1.11)$$

where a is the closest distance of approach of the centre of a neighbouring small ion to the centre of the sphere, and is usually taken as $a = b + 2.5$ Å. k is the Boltzmann constant.

The electrical free-energy change for ionization of the carboxyl group at ionic strength I is given by

$$(\varDelta G°)_I = \frac{Ne^2}{2D_0} \left(\frac{1}{b_1} - \frac{\varkappa}{1+\varkappa a_1} \right) + \frac{Ne^2}{2D_0} \left(\frac{1}{b_2} - \frac{\varkappa}{1+\varkappa a_2} \right), \qquad (1.12)$$

where b_1 and b_2 are the radii of a proton and the carboxylate ion, respectively.

The difference between the electrical free-energy change at ionic strength I and that at another ionic strength I' is given by

$$(\varDelta G°)_{I'} - (\varDelta G°)_I = \frac{Ne^2}{2D_0} \left(\frac{\varkappa}{1+\varkappa a_1} - \frac{\varkappa'}{1+\varkappa' a_1} + \frac{\varkappa}{1+\varkappa a_2} - \frac{\varkappa'}{1+\varkappa' a_2} \right). \qquad (1.13)$$

When I' is larger than I, the value of $(\varDelta G°)_{I'} - (\varDelta G°)_I$ is negative. Thus it is expected that the increase in ionic strength will decrease the pK_a value.

In the case of the protein molecule, many charges are located on the surface, and the ionization of an ionizable group is strongly affected by local charges. Such an electrostatic effect is also seen in the ionization of amino acids. The pK_a value of the carboxyl group of glycine $(NH_3^+ \cdot CH_2 \cdot COOH)$ is 2.35, and that of the carboxyl group of propionic acid (CH_3CH_2COOH), which has the same electronic structure as glycine, is 4.87. Thus the acidity of the carboxyl group of glycine is about 300 times higher than that of propionic acid. In the pH range for ionization of the carboxyl group, the amino group of glycine has the form NH_3^+, and this positive charge repels the proton from the carboxyl group. Such repulsion does not occur for the ionization of the carboxyl group of propionic acid. Therefore, the carboxyl group of glycine is ionized more easily than that of propionic acid.

The pK_a value of an ionizable group changes when it is linked by a hydrogen bond to another group. When the acidic form of an acid forms a hydrogen bond, the release of the proton is suppressed and the pK_a value increases. On the other hand, when the conjugate base of an acid is the acceptor of a hydrogen bond, it does not accept a proton readily and thus the pK_a value decreases.

As described above, many factors affect ionization behaviour. Each ionizable group in a protein molecule is located in a different environment and thus it is difficult to determine the pK_a value of any individual ionizable group. Some ionizable groups will be hydrogen-bonded and others will be located near charged groups. Furthermore, each ionizable group will be located in an environment with a different dielectric constant. Ionizable groups play an important role in protein function such as enzymatic action, and thus it is very important to understand the ionization behaviour of these groups even though this may be very difficult to determine. This subject will be dealt with in Chapters 4 and 8.

1-3. HYDROPHILIC AND HYDROPHOBIC AMINO ACID RESIDUES

Asn and Gln, which have an amide group, and Ser and Thr, which have a hydroxyl group in their side chains are all hydrophilic, as are other residues with similar ionizable groups. On the other hand, Ala, Val, Leu, Met, Pro Phe and Trp, the side chains of which have no affinity for water, are hydrophobic. We will first treat these hydrophilic and hydrophobic properties quantitatively.

The chemical potential ($\mu_{i,w}$) of a solute i dissolved in water (w), when the concentration of the solute is expressed by a mole fraction ($X_{i,w}$), is given by the following equation

$$\mu_{i,w} = \mu_{i,w}^{\circ} + RT \ln X_{i,w} + RT \ln \gamma_{i,w}, \tag{1.14}$$

where $\gamma_{i,w}$ is the activity coefficient based on the mole fraction.

The thermodynamic quantities obtained using the mole fraction as the concentration unit are known as quantities expressed in unitary units. When the concentration of a solute is expressed in terms of the mole fraction, the term $RT \cdot \ln X_{i,w}$ corresponds to the contribution of the mixing entropy of a solvent and solute to the chemical potential (cratic term), and thus the standard chemical potential $\mu_{i,w}^{\circ}$ contains only the free energy of the solute molecule itself and the free energy of interaction between the solute and solvent. If another concentration unit (for instance, molar concentration) is used, the contribution of mixing entropy to the chemical potential is contained in $\mu_{i,w}^{\circ}$.

The chemical potential of a solute i dissolved in another solvent A can be expressed by the same equation

$$\mu_{i,A} = \mu_{i,A}^{\circ} + RT \ln X_{i,A} + RT \ln \gamma_{i,A}. \tag{1.15}$$

If the chemical potentials of solute i as saturated solutions in water and in solvent A are expressed by $\mu_{i,w}$ and $\mu_{i,A}$, respectively, the following relation holds

$$\mu_{i,w} = \mu_{i,A} = \mu_{i(\text{crystal})}. \tag{1.16}$$

Thus

$$\mu_{i,w}^{\circ} + RT \ln X_{i,w} + RT \ln \gamma_{i,w} = \mu_{i,A}^{\circ} + RT \ln X_{i,A} + RT \ln \gamma_{i,A}. \tag{1.17}$$

The free energy of the solute itself is considered to be independent of the solvent used, and thus $(\mu_{i,w}^{\circ} - \mu_{i,A}^{\circ})$ represents the difference between the chemical potential of the interaction between solute i and water and that of the interaction between solute i and solvent A. Consequently, we can define this term as the free energy change (ΔG_t) of transfer of solute i from solvent A to water. ΔG_t is given by the equation

$$\Delta G_t = RT \ln \frac{X_{i,A}}{X_{i,w}} + RT \ln \frac{\gamma_{i,A}}{\gamma_{i,w}}, \tag{1.18}$$

and thus ΔG_t can be determined by measuring the solubilities of solute i in water and in solvent A.*

Nozaki and Tanford (1971) measured the solubilities of various amino acids in water and nonpolar solvents such as ethanol and dioxane, and determined the free-energy change of transfer from a nonpolar solvent to water using equation (1.18). By subtracting the transfer free-energy change for glycine $\overset{\text{H}_3^+\text{N}-\text{CH}-\text{COO}^-}{\underset{\text{H}}{\mid}}$ from that for any other amino acid $\overset{\text{H}_3^+\text{N}-\text{CH}-\text{COO}^-}{\underset{\text{R}}{\mid}}$, the transfer free-energy change for the side chain R can be estimated. The transfer free-energy change for the side chain is expressed by Δg_t. For this procedure to be valid, there must be additivity of the constituent parts of the whole amino acid molecule, and ΔG_t must be represented by the sum of Δg_t for each part. Comparisons of the transfer free energies between ethane and methane, alanine and glycine, and leucine and valine all give a value of about 730 cal as the value of the transfer free energy of CH_3. This indicates that additivity holds for the transfer free energy of the amino acid molecule. The value

---

*When enzyme E and substrate S bind in a ratio of 1 : 1 to form a complex, ES, the free energy change for the reaction

$$E+S=ES$$

can be expressed by

$$\Delta G° = -RT \ln K,$$

where K is the equilibrium constant ($K=(ES)/(E)(S)$). The standard free-energy change ($\Delta G°$) strictly represents the free-energy change when the reactants E and S and product ES are all 1.0 M (unit activity). In this reaction, however, one molecule of the product is formed from two reactant molecules, and thus the reaction accompanies a change in the number of molecules. Therefore, the mixing free-energy change for the left side of the reaction and that for the right side are different, and this difference is contained in $\Delta G°$. This term is different from the free-energy change of interaction between E and S only. For the reaction in a dilute solution, the mixing free-energy change (ΔG_{mix}) at 25°C is approximated by

$$\Delta G_{mix} \fallingdotseq RT \ln 55.5 \fallingdotseq 2,400 \text{ cal/mol}.$$

The free energy change (ΔG_u) only for the interaction between E and S is obtained by subtraction of ΔG_{mix} from the standard free-energy change

$$\Delta G_u = \Delta G° - 2,400 \text{ (cal/mol)},$$

where ΔG_u corresponds to the unitary part of the standard free-energy change. ΔG_u is also obtained directly using the equilibrium constant obtained in terms of the mole fraction unit of E, S, and ES.

TABLE 1-4
The Hydrophobicities of the Side Chains of Amino Acid Residues

Side chain	Δg_t (kcal/mol)		Side chain	Δg_t (kcal/mol)	
	(1)	(2)		(1)	(2)
Glycine	0	0	1/2 Cystine	+1.0	+1.3
Alanine	+0.5	+0.4	Serine	−0.3	−0.1
Valine	+1.5	+1.7	Threonine	+0.4	+0.4
Leucine	+1.8	+2.3	Asparagine	0.0	−0.8
Isoleucine	+3.0	+2.5	Glutamine	−0.1	−0.3
Phenylalanine	+2.5	+2.4	Aspartic acid	+0.5	−1.1
Tyrosine	+2.3	+1.3	Glutamic acid	+0.5	−0.9
Tryptophan	+3.4	+3.1	Histidine	+0.5	+0.2
Proline	+2.6	+1.0	Lysine	—	−1.4
Methionine	+1.3	+1.7	Arginine	—	−1.4
Cysteine	—	+2.1			

(1) Tanford, C. (1962) *J. Am. Chem. Soc., 84,* 4240; Nozaki, Y. & Tanford, C. (1971) *J. Biol. Chem., 246,* 2211.
(2) Fauchère, J.-L. & Pliška, V. (1983) *Eur. J. Med. Chem. Chim. Therm., 18,* 369.

of Δg_t is not sensitive to the nonpolar solvent used. The Δg_t values thus obtained are called the hydrophobicity of the side chains, and are given in Table 1-4.

The transfer free energy of the side chain can also be determined by measurement of the partition coefficient between water and a non-polar solvent. Fauchère and Pliška (1983) measured the partition coefficients (D) between water at neutral pH and octanol, of acetylaminoacidamides $CH_3CONHCHCONH_2$, instead of free amino
$\quad\quad\quad\quad\quad\quad\quad\quad\quad\quad\quad\quad\quad\quad\quad\quad\quad\quad\quad\quad\quad R$
acids, because the ionization states of the carboxyl and amino groups of free amino acids may affect the partition, and determined the free energy change of transfer of the side chain as follows

$$\Delta g_t = -RT\pi \qquad\qquad (1.19)$$

where

$$\pi \text{ (side chain)} = \log D \text{ (acetylaminoacidamide)}$$
$$-\log D \text{ (acetylglycineamide)}$$

The values of Δg_t thus obtained are also given in Table 1-4. The

Δg_t values obtained by measurement of partition coefficients are in good agreement with those obtained from solubilities.

Amino acid residues with large positive Δg_t values are hydrophobic. They will avoid contact with water and have a tendency to assemble in the interior of the protein molecule. On the other hand, amino acid residues with negative Δg_t values or small positive values are hydrophilic, and will have a tendency to be located on the surface of the protein molecule. The Δg_t values are thus very important for discussing the conformation and distribution of amino acid residues and the stability of protein molecules.

1-4. AROMATIC AMINO ACID RESIDUES

The aromatic amino acid residues, Phe, Tyr and Trp, absorb light in the ultraviolet region, and ultraviolet absorption spectroscopy is frequently used to determine protein concentration or for examining the environments of the residues. Table 1-5 shows the wavelengths at the absorption maxima and molar extinction coefficients of phenylalanine, tyrosine, and tryptophan. The ultraviolet absorption spectrum of a protein is changed by additives such as organic solvents, salts or substrates, and by variations in pH or temperature. However, since these changes are generally very small, the difference spectral technique is usually employed to detect them. This method consists of direct measurement of the absorption of a protein solution in a sample cell against a reference protein solution at exactly the same concentration. The difference spectra of tryptophan or tryptophyl residues in proteins caused by environ-

TABLE 1-5
Ultraviolet Absorbance of Aromatic Amino Acids

Amino acid	Wavelength at absorbance maximum[a] (nm)	Molar extinction coefficient (l/cm/mol)
Tryptophan	278	5,500
Tyrosine	275	1,340
Phenylalanine	257	190

[a]Absorbance maximum at the longest wavelength.

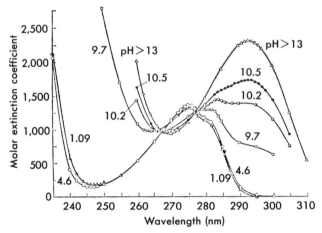

Fig. 1-2. Ultraviolet absorption spectra of glycyl-L-tyrosine. 25°C, ionic strength 0.16. (Edsall, J.T. & Wyman, J. (1958) "Biophysical Chemistry", Vol. 1, Academic Press, New York, p. 427)

mental perturbants exhibit positive or negative peaks at around 275, 285 and 293 nm. In the case of tyrosine or tyrosyl residues, positive or negative peaks appear at around 275 and 285 nm. Phenylalanine contributes little or nothing in the absorption region above 270 nm. The states of the tryptophyl residues of a protein can thus be examined using the peak at around 293 nm. Qualitatively, the smaller the dielectric constant of the medium, the longer the wavelength at which the peak appears and the larger the difference in the molar extinction coefficient.

The ultraviolet absorption spectrum changes greatly when the phenolic hydroxyl group of tyrosine is ionized. The titration curve of tyrosine can thus be determined by measuring the change in absorption with pH. This technique is called spectrophotometric titration. As an example, the ultraviolet absorption spectra of glycyl-L-tyrosine at various pH values are shown in Fig. 1-2. It can be seen that isosbestic points at which the absorption does not change appear at 267 and 278 nm. This indicates that the following equilibrium holds

TABLE 1-6
Fluorescence of Aromatic Amino Acids

Amino acid	Wavelength at fluorescence maximum (nm)	Quantum yields		
		Teale & Weber (1957)	Chen (1967)	Børresen (1967)
Tryptophan	348	0.20	0.13	0.119
Tyrosine	304	0.21	0.14	0.088
Phenylalanine	282	0.04	0.024	—

Teale, F.W. & Weber, G. (1957) *Biochem. J., 65,* 476. Chen, R.F. (1967) *Anal. Lett., 1,* 35. Børresen, H.C. (1967) *Acta Chem. Scand., 21,* 920.

TABLE 1-7
The Wavelengths at the Fluorescence Maxima of N-Acetyl-tryptophanmethylester in Various Solvents

Solvent	Dielectric constant	Wavelength at fluorescence maximum (nm)
Water	78.5	350
Methanol	32.6	340
n-Propanol	20.1	340
n-Butanol	17.1	340
Butylether	(4)	330
Dioxane	2.2	330
n-Hexane	1.9	310

Cowgill, R.W. (1967) *Biochim. Biophys. Acta, 133,* 6.

$$-CH_2-\langle\bigcirc\rangle-OH \rightleftharpoons -CH_2-\langle\bigcirc\rangle-O^- + H^+ .$$

The molar extinction coefficient of tyrosine at 295 nm in the pH region where the phenolic hydroxyl group is not ionized is zero. At pH values above 13 where the ionization of the phenolic hydroxyl group is complete, the molar extinction coefficient at 295 nm is 2,325 l/cm/mol. On ionization of the phenolic hydroxyl group, the absorption at 245 nm also increases to about 10,000 l/cm/mol.

In addition to the aromatic amino acid residues, the peptide bond absorbs light below 230 nm and the disulphide bond absorbs light weakly at around 250 nm.

Fluorescence measurements are also used to examine the states of aromatic amino acid residues. The fluorescence of simple proteins is due mainly to Trp and Tyr, and the contribution of Phe is small. For proteins which contain both Trp and Tyr, the contribution of the fluorescence of Trp is larger than that of Tyr. Table 1-6 shows the wavelengths at the fluorescence maxima and the quantum yields of tryptophan, tyrosine and phenylalanine when measured with 280 nm light for excitation.*

The wavelength at the fluorescence maximum is dependent on the polarity of the solvent. Table 1-7 shows the maximum wavelengths of N-acetyltryptophanmethylester in various solvents. As can be seen, as the polarity of the solvent decreases, the maximum wavelength decreases from 350 nm. It is also known that the unionized carboxyl group and the disulphide bond quench the fluorescence of Trp. On the basis of these characteristics, the environment of Trp in the protein molecule can be investigated experimentally.

---

*For observation of the fluorescence of Trp only, 295 nm light is preferable for excitation.

2

Secondary Structure of Proteins

2-1. BASIC STRUCTURES OF POLYPEPTIDE MAIN CHAINS

Chapter 1 described the physicochemical properties of the side chains of amino acid residues. This chapter deals with the basic structures of the polypeptide main chains which are common to all protein molecules. The term secondary structure refers to the conformations of amino acid residues which are close to each other in the primary structure. The bond length and bond angles of the peptide bond are shown in Fig. 2-1. The C–N bonds in the –C–N– groups have a partial double bond character due to resonance between the structures

$$C_\alpha\text{–}\underset{\underset{O}{\|}}{C'}\text{–}\overset{\overset{H}{|}}{N}\text{–}C_\alpha \quad \text{and} \quad C_\alpha\text{–}\underset{\underset{O^\ominus}{|}}{C'}\text{=}\overset{\overset{H}{|}}{N^\oplus}\text{–}C_{\alpha'}.$$

In simple peptides, it has been shown that the bond length (1.24 Å) of C′–O is shorter than the bond length (1.43 Å) of C–O in ether,

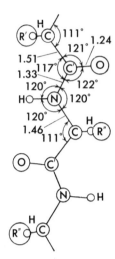

Fig. 2-1.   Bond distances (Å) and angles in a peptide unit.

and that the bond length (1.33 Å) of C′–N is shorter than the bond length of C–N, but longer than the length of the double bond. These facts confirm the double bond character of the peptide bond.

Free rotation about the C′–N bond is impossible owing to its double bond character, and the six atoms lie in the same plane. There are two possible configurations, *cis* and *trans*, of a coplanar structure. They differ in the arrangement of the C_α group with respect to their non-rotating C′–N bonds (Fig. 2-2). However, the *trans* form is about 3 kcal/mol more stable than the *cis* form, and occurrence of the *trans* form is far more common in most free polypeptide chains. The probability of a peptide bond occurring in the *cis* form is only about 10^{-3}. However, the peptide bond, X-Pro, where X is Pro or any other residue, has a probability of 0.1–0.3 of occurring in the *cis* form (Fig. 2-2).

X-Pro bonds of *cis* form are found in some proteins: for instance, -Pro 93 and -Pro 114 in ribonuclease S, -Pro 168 in sub-tilisin, -Pro 29 and -Pro 200 in carbonic anhydrase, and -Pro 8, -Pro 95 and -Pro 148 in Bence-Jones protein. Peptide bonds which have no Pro on the C-terminal side but have a *cis* form occur in Ser

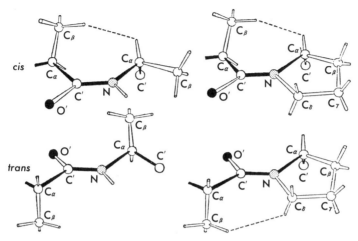

Fig. 2-2. *Cis* and *trans* arrangements of Ala-Ala (left) and Ala-Pro (right). The dotted lines indicate unfavourable interactions between the hydrogen atom at C_β and the hydrogen atom at C_α of the next residue. This unfavourable interaction exists in *cis*, but not in *trans* Ala-Ala. In the case of Ala-Pro, this interaction exists in both the *cis* and *trans* arrangements. (Blundell, T.L. & Johnson, L.N. (1976) "Protein Crystallography," Academic Press, London, p. 27)

197-Tyr 198, Pro 205-Tyr 206 and Arg 272-Asn 273 in carboxy-peptidase A.

It might be expected that the polypeptide chain could assume many conformations if free rotation was possible around each bond in the chain. In fact, however, there are some restrictions. One is the restriction of rotation around the C'-N bond due to its partial double bond character. The only bonds which can rotate freely are the N-C_α bond and C_α-C' bond. The rotations around the N-C_α and C_α-C' bonds are defined by dihedral angles ϕ (phi) and ψ (psi), respectively, and the folding of the polypeptide chain backbone can be expressed using these angles (Fig. 2-3). When ϕ and ψ are changed keeping the angle N-C_α-C' fixed at the tetrahedral angle (111°), collisions will occur between the atoms of side chains and main chain. This restricts the rotation around bonds N-C_α and C_α-C' to some extent. The extent of these collisions between atoms depends on the type of side chains present, and is different for side chains of different lengths. For longer side chains which extend to

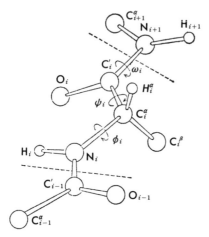

Fig. 2-3. Dihedral angles in a polypeptide. When the peptide unit rotates clockwise (viewed from C_α) around the C-N and C-C' bonds while the C_α is held fixed, the angles are taken to be positive.

TABLE 2-1
Non-bonded Contact Distances

	Normal limit (Å)	Extreme limit (Å)
H ⋯ H	2.00	1.90
H ⋯ O	2.40	2.20
H ⋯ N	2.40	2.20
H ⋯ C	2.40	2.20
O ⋯ N	2.70	2.60
O ⋯ C	2.80	2.70
N ⋯ N	2.70	2.60
N ⋯ C	2.90	2.80
C ⋯ C	3.20	3.00

Ramachandran, G.N. & Sasisekharan, V. (1968) *Adv. Protein Chem., 23*, 283.

C_β and/or C_γ, however, there will be conformations in which no collisions occur between non-bonded atoms because of the rotation around C_α-C_β, and the extent of the collision will be small for atoms beyond C_γ. Allowed conformations in which all the interactions between atoms are acceptable, and "partially allowed" conformations in which unfavourable repulsion between certain atoms

is overcome by other attractive effects, have been given by Ramachandran and Sasisekharan (1968) (Table 2-1 and Fig. 2-5). The
atoms of each amino acid residue in the polypeptide chain must
have ϕ and ψ values in these acceptable or partially acceptable
regions (Gly is the exception).

As described below, regular arrangements of the linear polypeptide chain include the α-helix, β-structure, reverse turn (also
called bend or β-turn) and polyproline-type helix. Structures other
than these regular structures are sometimes called disordered structure.* However, this disordered structure in proteins is quite different from the random coil of polymers, and simply means that it has
no repeating regular structure such as the α-helix, but has well
defined arrangements of residues with definite ϕ and ψ values.

A) α-Helix

On the basis of structural chemical studies on amino acids and
simple peptides, Pauling *et al.* (1951) investigated the conformations of the polypeptide chain which satisfy the coplanarity, bond
length, and bond angle of the peptide group and the capability of
forming maximal intrachain hydrogen bonds between $C'=O$ and
H–N groups. The hydrogen bond is a type of electrostatic interaction formed between a hydrogen atom which has a partial positive
charge and highly electronegative atoms such as nitrogen and
oxygen with partial negative charges. Since, as described above, the
N–H and $C'=O$ groups contain partial positive and negative
charges, they are suitable for forming a hydrogen bond

$$\overset{\delta-\ \delta+\ \ \delta-}{>N–H\cdots O=C<.}$$

Pauling *et al.* (1951) and Pauling and Corey (1951) proposed
two principal models for the arrangement of the polypeptide chain,
both of which form maximal hydrogen bonds between the N–H
and C=O groups of the peptide bond, keeping the geometry of the

---

*This is different from the meaning of the word "disordered" used by crystallographers.
They use "disordered" when atomic locations are not well defined.

peptide bond as it is in simple peptides. One is the α-helix, which has hydrogen bonds between N–H and C=O groups nearby in the linear sequence, and the other is the β-structure, which has inter- or intrapeptide hydrogen bonds between distant N–H and C=O groups. The α-helix model proposed by Pauling *et al.* was confirmed in myoglobin by X-ray crystallographic analysis ten years later by Kendrew *et al.* (1960).

The α-helix has the following properties (see Fig. 2-4).

(i)   The hydrogen bond is formed between the N–H group of an amino acid residue n and the carbonyl oxygen of residue $(n-4)$. The ring formed by a single hydrogen bond contains 13 atoms.* A single turn of the helix contains 3.6 residues and thus the α-helix can also be called the 3.6_{13}-helix.

(ii)   A residue in the helix extends 1.5 Å and thus a single

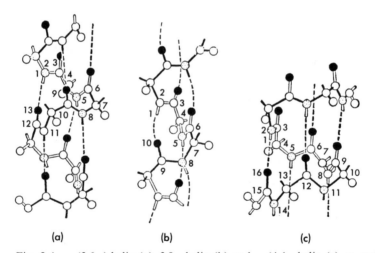

(a)   (b)   (c)

Fig. 2-4. $\alpha(3.6_{13})$-helix (a), 3.0_{10}-helix (b) and π $(4.4_{16}$-helix (c). $\circ$, carbon; $\odot$, nitrogen; $\bullet$, oxygen. (Blundell, T.L. & Johnson, L.N. (1976) "Protein Crystallography," Academic Press, London, p. 32)

---

*In the ring formed by a hydrogen bond

$$\begin{array}{c} \text{O} \text{-} \text{-} \text{-} \text{-} \text{-} \text{-} \text{-} \text{-} \text{-} \text{H} \\ \| \qquad\qquad\qquad | \\ \text{-C-(NH-CH-CO)}_n\text{-N-} \end{array}$$

there are $N = 3n + 4$ atoms. For 3.0_{10} helix, $n = 2$ and $N = 10$, for α-helix $n = 3$ and $N = 13$ and for π-helix $n = 4$ and $N = 16$.

turn of the helix extends 5.4 Å (1.5 Å × 3.6) along the long axis.

In the α-helix, the C=O bond is parallel to the helix axis and a straight hydrogen bond is formed with the N-H group. Therefore, the hydrogen bond between the N-H and C=O groups in the α-helix forms the most stable geometrical arrangement. When the hydrogen bond $N^- - H^+ \cdots O^-$ is not straight, the partial negative charge on the nitrogen atom repels the partial negative charge on the oxygen atom and the hydrogen bond is less stable. The pitch of the helix can be right-handed or left-handed, but a right-handed helix is more stable than a left-handed one, because the β-carbon and carbonyl oxygen are too close in the left-handed helix.

The right-handed α-helix is stabilized not only by the straight hydrogen bonds between N-H and C=O groups, but also by the interactions of all the constituent atoms of the main chain that are packed closely together. The ϕ and ψ values of the α-helix (Table 2-2) are strictly confined to the regions in which no atoms collide, and small changes in ϕ and ψ values lead to unfavourable contacts. This can be easily understood from the plot of ψ against ϕ (Ramachandran plot, Fig. 2-5). In the α-helix, all the atoms are

TABLE 2-2
Basic Conformations of Polypeptide Chains Consisting of L-Amino Acids

Regular conformation	$\phi(°)$	$\psi(°)$	$\omega(°)$	Pitch of helix	Number of residues per pitch
Completely extended chain (E)	+180	+180	+180	7.3	2.00
Right-handed α-helix (α_R)	−57	−47	+180	5.4	3.62
Left-handed α-helix (α_L)	+57	+47	+180	5.4	3.62
3.0_{10}-helix (3_{10})	−49	−26	+180	6.0	3.0
π-helix (π)	−57	−70	+180	5.1	4.40
Parallel β-structure (β_P)	−119	+113	+180	6.5	2.0
Antiparallel β-structure (β_{AP})	−139	+135	−178	7.0	2.0
Right-handed polyglycine II (G_R)	+80	−150	+180	9.3	3.0
Left-handed polyglycine II (G_L)	−80	+150	+180	9.3	3.0
Right-handed poly-L-prolin (P_I)	−83	+158	0	5.9	3.33
Left-handed poly-L-prolin (P_{II})	−78	+149	+180	9.3	3.00
Collagen (C)	−51	+153			
	−76	+127	+180	9.5	3.3
	−45	+148			

ω, rotational angles about the C'-N bond, ϕ, ψ and ω, are 0° for the *cis* conformation.

closely packed and the helix is stabilized by their van der Waals interactions, in addition to hydrogen bonding.

In addition to the α-helix, the 3.0_{10}-helix and π-helix $(4.4_{16}$-helix) also exist as helical conformations of the polypeptide chain (Fig. 2-4). The 3.0_{10}-helix has three residues per turn and hydrogen bonds complete rings of 10 atoms. The π-helix has 4.4 residues per turn and hydrogen bonds complete rings of 16 atoms. These helices are also found in proteins, but not for more than a few residues.

Among the fibrous proteins, the α-helical content of keratin, myosin, epidermin and fibrinogen (k-m-e-f group) is very high. Keratin is present in hair and wool. Fibres made of α-helical proteins have some degree of elasticity and can recover from a small extension along the fibre axis, but cannot recover when the fibre is extended to such a degree that the hydrogen bonds of the helix are pulled apart. Such a change was observed for keratin by Astbury and Woods (1930), who called the two types α-keratin and β-keratin. α-Keratin corresponds to the α-helix and β-keratin corresponds to the β-structure described below.

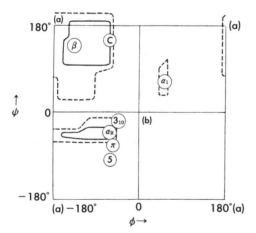

Fig. 2-5. Plots of dihedral angles in polypeptides (Ramachandran plot). The solid lines enclose fully-aligned regions and the broken lines indicate outer limits. α_R, right-handed α-helix; α_L, left-handed α-helix; 3_{10}, 3_{10}-helix; π, π-helix; β, β-structure; C, collagen-helix; 5, the conformation in a five-membered planar ring.

B) β-Structure

A polypeptide chain which forms β-structure is somewhat extended, although not completely so, and pleats are formed as shown in Fig. 2-6. A β-structure sheet which is formed by two or more polypeptide chains is called a β-pleated sheet. A single polypeptide chain in a β-pleated sheet is called a β-strand. The hydrogen-bonded C=O and N–H groups of the β-strands protrude at right angles to the polypeptide chains and are situated at positions that are favourable for formation of hydrogen bonds. Both interchain and intrachain β-structures exist. The interchain β-structure is found in fibrous proteins such as silk fibroin, and the intrachain β-structure is present in many globular proteins. There are two types of β-sheet; one is a parallel β-sheet in which the directions of the β-strands run parallel, and the other is an antiparallel β-sheet in which the directions of the β-strands are antiparallel (Fig. 2-7). Hydrogen bond networks are different in parallel and antiparallel sheets. The ϕ and ψ angles of the parallel and antiparallel β-sheets are given in Table 2-2 and Fig. 2-5. The antiparallel β-sheet was first found in hen egg-white lysozyme by X-ray crystallographic analysis (Blake *et al.*, 1965).

In silk fibroin, the β-sheets are packed together in layers, and the side chains of the β-strands, which protrude alternately above and below the plane of the sheet, are packed together efficiently. For the packing to be complete, the side chains must be short. Silk

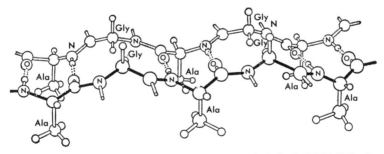

Fig. 2-6.   Antiparallel β-sheet. (Dickerson, R.I. & Geis, I. (1969) "The Structure and Action of Proteins," W.A. Benjamin, Inc. Menlo Park, Calif., p. 36)

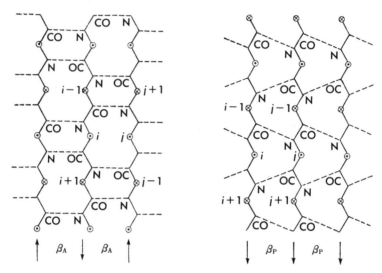

Fig. 2-7. Antiparallel (left) and parallel (right) β-sheet. The dotted lines indicate hydrogen bonds. The arrows show chain direction. ⊙, upward side chains; ⊗, downward side chains. (Schulz, G.E. & Schirmer, R.H. (1979) "Principle of Protein Structure," Springer-Verlag, New York, p. 76)

fibroin has a high content of Gly, Ala and Ser with short side chains, and has regions of a repeating sequence, (-Gly-Ser-Gly-Ala-Gly-Ala-)$_n$. Silk fibroin has also a low content of Tyr, Arg, Val, Asp and Glu, which have bulky side chains. Where they do occur, the regular layered structures are disrupted to form unordered structures.

Whereas fibres made of α-helix can be extended along the fibre axis to an extent allowed by the hydrogen bonds, fibres made of β-sheets cannot be extended along the fibre axis without disrupting the covalent bonds.

C) Reverse Turn

In the reverse turn conformation, a hydrogen bond is formed between the carbonyl oxygen of amino acid residue n and the N–H group of the residue $(n+3)$. Two typical types of reverse turn are shown in Fig. 2-8. In type I, the carbonyl group of the middle

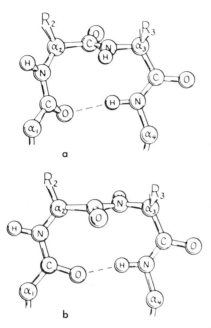

Fig. 2-8. Reverse-turn conformations. (a) type I. (b) type II. (Richardson, J.S. & Richardson, D.C. (1989) in "Prediction of Protein Structure and the Principles of Protein Conformation," ed. by Fasman G.D., Plenum Press, New York, p. 24)

amide plane points in the opposite direction from the side chains of both adjacent residues. In type II, it points in the same direction, and thus for steric reasons the third residue must be Gly. The type I reverse turn has $\phi_2 = -60°$, $\psi_2 = -30°$, $\phi_3 = -90°$ and $\psi_3 = 0°$, and type II has $\phi_2 = -60°$, $\psi_2 = -120°$, $\phi_3 = +80°$ and $\psi_3 = 0°$. The conformation of the type III reverse turn corresponds to that of one turn of the 3.0_{10}-helix. The reverse turn is a very important conformation in that it reverses the direction of the polypeptide chain by 180°.

D) Polyproline-type Helix

Poly-L-proline assumes the right-handed type I helix immediately upon dissolution in acetic acid, and then gradually changes to the left-handed helix, which has three residues per turn (Fig. 2-9).

A typical example of this helix is found in collagen, which is constituted mainly from Gly and Pro. In this helix, the C=O and N–H groups point almost at right angles to the helix axis, and no hydrogen bonds are formed between them within the same chain. The proline residue has no hydrogen at the nitrogen atom and thus cannot form a hydrogen bond. In the triple helix observed in collagen, however, the proline residue can be accommodated well in the helix. In the polyglycine helix, the N–H and C=O groups which point at right angles to the helix axis can form hydrogen bonds between the C=O and N–H groups, respectively, of the adjacent chain.

The triple polyproline-type helix is the basic structure of collagen, which is found in connective tissue (Fig. 2-10). In collagen, the N–H groups of every third amino acid residue of the three chains form hydrogen bonds with C=O groups of one of the other two chains. The C_α atom of this residue is located near the axis of the triple helix, and this must be Gly since there is no room to accept bulky side chains. The side chains and NH groups of the other two residues in the repeating unit of each chain are more distant from the helix axis, and thus these residues can have any kind of side chain. Therefore, a repeating sequence such as (-Gly-X-Pro-)$_n$, (-Gly-X-Hypro*-)$_n$ or (-Gly-Pro-Hypro-)$_n$ is suitable for formation of the triple helix of collagen, where X is any other

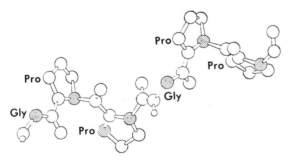

Fig. 2-9. Polyproline-type helix of (-Gly-Pro-Pro-)$_n$ (Fletterick, R.J., Schroer, T. & Matela, R.J. (1985) "Molecular Structure. Macromolecules in Three Dimensions," Blackwell Sci. Publ., Oxford, p. 72)

---

*Hypro: hydroxyproline.

amino acid residue. In fact, the content of Gly in collagen is about 33 mole per cent. The collagen fibre is strongly constructed, because it consists of three helical chains connected by hydrogen bonds. Therefore, the collagen triple helix is utilized when strong materials such as tendon, cowhide or bladder are needed. Short polyproline-type helical conformations are also found in globular proteins such as the hinge region of immunoglobulins, pancreatic polypeptide and bovine pancreatic trypsin inhibitor. The amino acid sequence

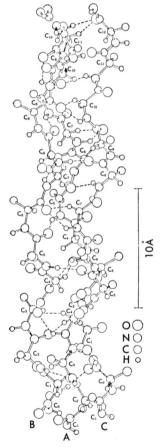

Fig. 2-10. Collagen helix. (Ramachandran, G.N. (1967) in "Treatise on Collagen," Vol. 1, ed. by G.N. Ramachandran, Academic Press, New York, p. 127)

2 4 6 8 10
H–Gly–Pro–Ser–Gln–Pro–Thr–Tyr–Pro–Gly–Asp–

12 14 16 18 20
Asp–Ala–Pro–Val–Glu–Asp–Leu–Ile–Arg–Phe–

22 24 26 28 30
Tyr–Asp–Asn–Leu–Gln–Gln–Tyr–Leu–Asn–Val–

32 34 36
Val–Thr–Arg–His–Arg–Tyr–NH_2

Fig. 2-11. Primary structure of chicken pancreatic polypeptide.

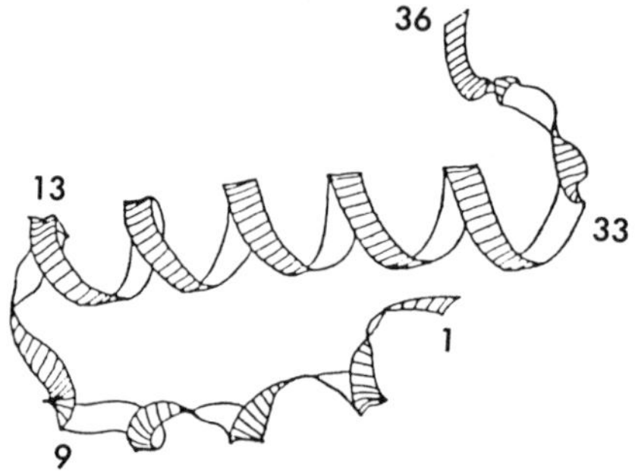

Fig. 2-12. Crystallographic structure represented by the ribbon model of chicken pancreatic polypeptide. (Glover, I. *et al.* (1983) *Biopolymers, 22*, 293)

of complement C1q suggests the presence of the polyproline-type helix. The amino acid sequence and crystallographic structure of chicken pancreatic polypeptide are shown in Figs. 2-11 and 2-12, respectively. In spite of the fact that it is a short polypeptide consisting of 36 residues, the pancreatic polypeptide molecule assumes a globular conformation and consists of a polyproline II-type helix (from the N-terminal residue to residue 8) and an α-helix (from residue 14 to residue 32) (Blundell *et al.*, 1981; Glover *et al.*, 1983). These are linked by a reverse turn (from residue 9 to residue 13) and are closely packed through hydrophobic interactions. In the region from the N-terminal residue to residue 8, the proline residues are found at positions 2, 5 and 8 and the average values of ϕ and ψ are $-70°$ and 138°, respectively.

As indicated by the ϕ and ψ values (Table 2-2), the left-handed polyglycine II-type helix is the same as the left-handed poly-L-proline II-type helix.

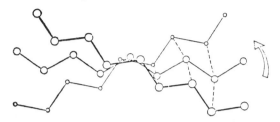

Fig. 2-13.   β-Twist. (Sternberg, M.J.E. & Thornton, J.M. (1978) *Nature, 271*, 15)

2-2.   SECONDARY STRUCTURE OF PROTEINS

A) Rules of Secondary Structure

The following tendencies have been observed for secondary structure in globular proteins.

1)   In globular proteins that have been analyzed by X-ray crystallography, about one quarter of the total amino acid residues are found in the α-helix. A single α-helix usually consists of 10–15 residues. Almost all of the helical conformations found in proteins are α-helices. 3.0_{10}-helices seldom exist, but are sometimes found at the carboxyl end of an α-helix. No 3.0_{10}-helix exists at the N-terminal end of an α-helix. In the 3.0_{10}-helix, a straight hydrogen bond is not formed between N–H and C=O groups, and thus the helix is not as stable as the α-helix. In the π-helix, there is a large space around the helix axis, and van der Waals interactions are not utilized to stabilize the helix, in spite of the straight hydrogen bond between N–H and C=O groups.

2)   Single β-strands are usually 3 to 10 residues long. Most β-pleated sheets consisting of several β-strands are not planar, but twisted. This conformation is called the β-twist. As shown in Fig. 2-13, β-sheets have a left-handed twist when viewed along the sheet plane perpendicular to the strands, and this left-handed twist corresponds to a right-handed rotation of carbonyl and amide groups.

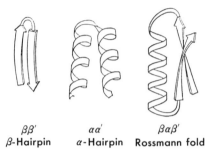

βββ'
β-Hairpin

αα'
α-Hairpin

βαβ'
Rossmann fold

Fig. 2-14. Basic supersecondary structures. (Janin, J. (1979) *Bull. Inst. Pasteur,* 77, 337)

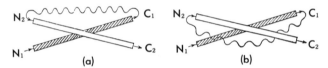

Fig. 2-15. Connection between two β-strands. (a) right-handed. (b) left-handed. (Sternberg, M.J.E. & Thornton, J.M. (1976) *J. Mol. Biol., 105,* 367)

3) Two α-helices form an α-hairpin and two β-strands form a β-hairpin. These assemblies of secondary structural elements are called supersecondary structure (Fig. 2-14). Another frequently observed supersecondary structure is two parallel β-strands of a β-sheet with a connection x between them, where x is an α-helix or another structure such as a non-regular chain. This $\beta x\beta'$ unit is also called the Rossmann fold after the name of its discoverer. The connection of the two β-strands in virtually all $\beta x\beta'$ units is right-handed (Fig. 2-15).

4) Most parallel β-sheets are protected by an α-helix or other structures. In the antiparallel β-sheet, on the other hand, one side of the sheet is exposed to solvent and the other is buried in the protein interior. Thus a hydrophilic residue and hydrophobic residue appear alternately in the amino acid sequence. The parallel β-sheet seems to be less stable than the antiparallel β-sheet due to the non-linear hydrogen bonds.

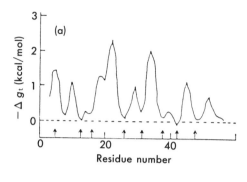

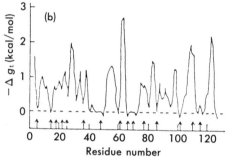

Fig. 2-16. The plots of hydrophobicity (Δg_t) against the amino acid residue number of bovine pancreatic trypsin inhibitor (a) and hen egg-white lysozyme (b). The arrows indicate the locations of reverse turns. (Jaenicke, R. (1987) *Prog. Biophys. Mol. Biol., 49,* 117)

5) The reverse turns are usually located on the surface of the protein molecule. This is clearly seen in the plot of the hydrophobicities (Δg_t) of the side chains (see Section 1-3) against the residue number (Fig. 2-16). As can be seen, the reverse turns are found at the minima of the Δg_t values.

B) Classification of Proteins on the Basis of Secondary Structure

Proteins may be classified into five groups on the basis of their secondary structure (Table 2-3). As will be described in Section 3-1, a large protein molecule usually consists of two or more smaller, structurally independent regions, called "domains." Table 2-3 shows the classification of the domain structures rather than the

TABLE 2-3
Classification of Proteins on the Basis of Secondary Structures

Classification	Secondary structures	Examples
α-Protein	With α-helix only	Myoglobin, haemoglobin, myogen, haemerythrin, bacteriorhodopsin, ferritin, phospholipase C
β-Protein	Mainly with β-sheet	Immunoglobulin, rubredoxine, superoxidodismutase, concanavalin A, prealbumin, chymotrypsin, bacteriochlorophyll protein
$\alpha+\beta$ Protein	With α-helix region and β-sheet region that exist apart in the sequence	Insulin, cytochrome b_5, pancreatic trypsin inhibitor, ribonuclease A, hen egg-white and T_4 phage lysozyme, papain, nuclease, thermolysin
α/β Protein	With alternating segments of α-helix and β-sheet	Thioredoxin, flavotoxin, alcohol dehydrogenase, D-glyceraldehyde-3-phosphate dehydrogenase, lactate dehydrogenase, subtilisin, adenylate kinase, hexokinase, phosphoglycerate mutase, carboxypeptidase A, triose phosphate isomerase, phosphoglycerate kinase, phosphorylase, Taka-amylase A
Coil protein	With no regular secondary structures	Bowman-Birk-type protease inhibitor

whole protein molecule. The secondary structures of proteins belonging to each class will be described below.

a) α-Proteins

The principal motif of α-proteins is the α-hairpin. The best known example is myoglobin. This protein was the first for which the secondary and tertiary structures were determined by X-ray crystallographic analysis. The myoglobin molecule contains eight helices (see Fig. 6-3). Of 153 amino acid residues, 121 belong to these α-helices and the helical content amounts to 79%. Haemoglobin has four subunits, and the structure of each subunit is very similar to that of myoglobin. Cytochrome c_{562}, cytochrome c', haemerythrin, ferritin and tobacco mosaic virus coat protein all have similar structures, in which four α-helices pack together in a bundle. However, the functions of these proteins are quite different. Phospholipase C obtained from *Bacillus cereus* consists of eight helices. This was the first enzyme found to consist of only α-helices (Hough *et al.*, 1989).

b) β-Proteins

Almost all β-proteins have the antiparallel β-structure and show the supersecondary structure of the $\beta\beta'$-hairpin. The secondary structures of immunoglobulins are described here as an example of β-proteins. As shown in Fig. 2-17, immunoglobulin G consists of two heavy (H) chains each with a molecular weight of about 50,000, and two light (L) chains each with a molecular weight of about 23,000. These chains interact by both noncovalent interactions and disulphide bonds. The antibodies obtained from the γ-globulin fraction of normal serum are heterogeneous with respect to their amino acid sequences. However, a single antibody protein (called myeloma protein) is synthesized in large amounts in patients suffering from multiple myeloma. The myeloma protein is very homogeneous and has the same four-chain structure as immunoglobulins. Furthermore, multiple myeloma patients excrete large

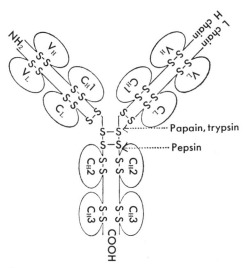

Fig. 2-17. Basic structure of human IgG1 immunoglobulin. The arrows indicate the cleavage points of pepsin, trypsin and papain.

amounts of a protein called Bence-Jones protein in urine. This protein has been found to correspond to the light chain of the myeloma protein. The amino acid sequences of Bence-Jones and myeloma proteins differ between patients, but the proteins obtained from any given patient are very homogeneous and are suitable for studies of primary, secondary and tertiary structures. Structural studies of immunoglobulins advanced considerably with the discovery of myeloma proteins. Today, monoclonal antibodies provide a source of homogeneous material for structural analysis.

The light and heavy chains of immunoglobulin G consist of independently folded regions (domains) each with about 120 amino acid residues. There are two domains (V_L and C_L) in the light chain and four (V_H, C_H1, C_H2 and C_H3) in the heavy chain. The amino acid sequences of the V_L and V_H domains (called variable regions) differ between antibody molecules of different specificity. In the variable regions there are segments of sequence in which variations are particularly marked, and these regions are called hypervariable regions. These hypervariable regions play an important role in determining the shape and structure of the antigen-binding site. The amino acid sequences of the other domains, C_L, C_H1, C_H2 and C_H3, are constant in all immunoglobulin G molecules, and are called the constant regions. These domains each have functionally important roles. The antigen binds to the region where the V_L and V_H domains assemble, and complement binds to the C_H2 domain.

It is interesting to note that all these domains have very similar secondary structures. Each domain consists of two β-sheets with one intrachain disulphide bond buried in the interior hydrophobic region between the sheets. Each β-sheet consists of three or four β-strands. The basic structure of the immunoglobulin domain is called the immunoglobulin-fold. The secondary structure of Bence-Jones protein Mcg is shown in Fig. 2-18. Despite the great variation in the amino acid sequence, the V_L and V_H domains also hold the basic immunoglobulin fold. The cell-surface glycoprotein CD4 (Wang et al., 1990; Ryu et al., 1990) and bacterial chaperon protein PapD (Holmgren & Branden, 1989) also have secondary structures very similar to the immunoglobulin fold. Figure 2-19 compares the secondary structures of the CD4, chaperon protein and immunoglobulin domains.

b) $\alpha + \beta$ Proteins

$\alpha + \beta$ proteins contain both α-helices and β-sheets, but these structures are distant from each other in the amino acid sequence. The β-sheets of most $\alpha + \beta$ proteins are antiparallel. The β-sheet was first discovered in the hen egg-white lysozyme molecule (Blake *et al.*, 1965). The structure of the hen lysozyme molecule is shown in Figs. 2-20 and 2-21. This lysozyme consists of 129 amino acid residues, and the helices are formed in the regions 5–15, 24–34, 80–

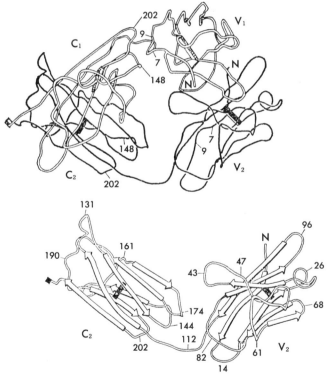

Fig. 2-18.   The secondary structure of Bence-Jones protein Mcg. The bold lines indicate the disulphide bonds. (Schiffer, M. *et al.* (1973) *Biochemistry 12*, 4620)

42

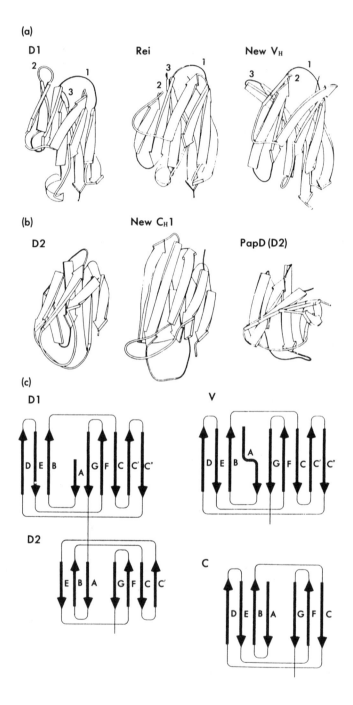

(a)

D1   Rei   New V_H
2 1   3 1   3 1
  3   2   2

(b)

        New C_H1
D2   PapD (D2)

(c)

D1   V
D E B A G F C C' C"   D E B A G F C C' C"

D2   C
E B A G F C C'   D E B A G F C

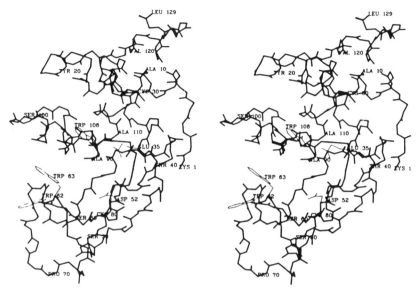

Fig. 2-20.   Stereo drawing of the structure of hen egg-white lysozyme. (Strynadka, N.C. & James, M.N.G. (1991) *J. Mol. Biol., 220*, 401)

85, 88–96 and 118–122; an antiparallel β-sheet is formed between the β-strand from 42 to 48 and the β-strand from 49 to 54. There is also a 3.0_{10}-helix of one turn from residue 119 to residue 122.

As other examples of $\alpha + \beta$ proteins, the structures of ribonuclease A, insulin and bovine pancreatic trypsin inhibitor are shown in Figs. 2-22, 2-23 and 2-24, respectively. Bovine pancreatic trypsin inhibitor is a very small protein consisting of 58 amino acid residues. There is an antiparallel β-sheet between the β-strand from 16 to 24 and the β-strand from 27 to 46, and a three-turn α-helix in the region 47–56. The physicochemical properties of ribonuclease A, hen lysozyme and pancreatic trypsin inhibitor have been studied in detail.

←Fig. 2-19.   Immunoglobulin-fold domains. D1 and D2 domains of the cell-surface glycoprotein CD4, the variable domain of light chain Rei, the variable (V_H) and constant (C_H1) domains of heavy chain New and D2 domain of the bacterial chaperon protein PapD. (Ryu, S.E. *et al.* (1990) *Nature 348*, 419)

44

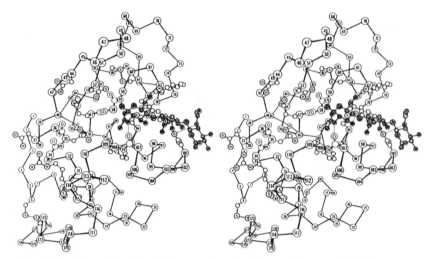

Fig. 2-21. Stereo drawing of the α-carbon backbone of hen egg-white lysozyme.
Binding of the trimer of N-acetylglucosamine (black circles) is also shown. (Dickerson, R.E. & Geis, I. (1969) "The Structure and Action of Proteins," W. Benjamin
Inc., Menlo Park, Calif., p. 72)

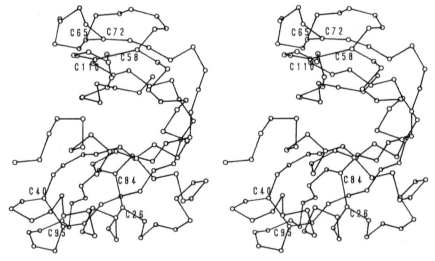

Fig. 2-22. Stereo drawing of the α-carbon backbone of bovine ribonuclease A.
Atomic coordinates for the drawing were obtained from Brookhaven Protein Data
Bank (entry set 3RN3). (Pilton, R.F. Jr. *et al.* (1992) *Biochemistry, 31,* 2469).
(Kindly provided by Dr. M. Sato)

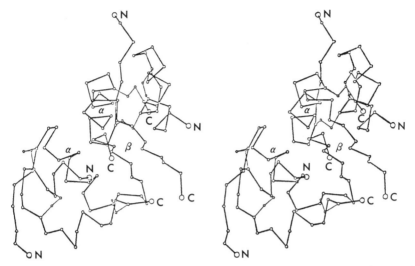

Fig. 2-23. Stereo drawing of the α-carbon backbone of insulin. N, amino terminus; C, carboxyl terminus; α, α-helix; β, intermolecular β-sheet. (Kindly provided by Dr. N. Sakabe)

d) α/β Proteins

The basic structure of proteins in this class is the supersecondary structure $\beta\alpha\beta'$. The β-sheets of most α/β proteins are parallel. As shown in Table 2-3, many proteins belong to this class. Almost all the glycolytic enzymes have this type of secondary structure.* Glyceraldehyde-3-phosphate dehydrogenase, alcohol dehydrogenase, lactate dehydrogenase and maleate dehydrogenase each consist of a catalytic domain and a coenzyme-binding domain. The coenzyme-binding domains of these dehydrogenases have very similar secondary structures, as shown in Fig. 2-25. The basic structural unit of these domains consists of six parallel β-strands

---

*Attempts to clarify the enzymatic reactions responsible for the 12 steps from glycogen to pyruvate in intracellular glycolysis at the atomic level were carried out through cooperation with the Universities of Bristol, Oxford, and Cambridge in the U.K. from the beginning of the 1970s, and valuable results were obtained, which were published in a book entitled "The Enzymes of Glycolysis: Structure, Activity and Evolution" (Royal Society of London, 1981).

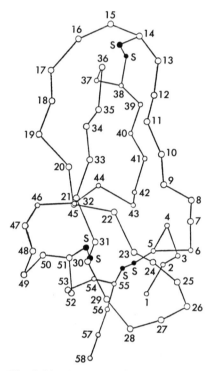

Fig. 2-24. Structure of bovine pancreatic trypsin inhibitor represented by the α-carbon backbone. (Huber, R. *et al.* (1970) *Naturwissenschaften, 57,* 389)

(βA, βB, βC, βD, βE and βF) and four α-helices (αB, αC, αE and α1F), and is expressed as

$$\beta \diagdown {}_{\alpha}\diagup \beta \diagdown {}_{\alpha}\diagup \beta \qquad \beta \diagup {}^{\alpha}\diagdown \beta \diagup {}^{\alpha}\diagdown \beta \; .$$

This is the structure connected by two units found in the FMN-binding part of the flavodoxin molecule.

In the subunit of triosephosphate isomerase, which consists of 248 amino acid residues, eight β-strands form a barrel-like β-sheet structure (called a β-barrel) (Figs. 2-26 and 2-27) and these β-strands are connected alternately by eight α-helices. This β-barrel is expressed as

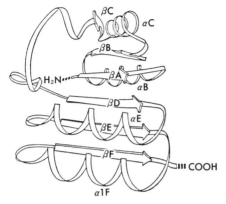

Fig. 2-25. The coenzyme-binding domain of dehydrogenase. (Matthews, B.W. (1976) *Annu. Rev. Phys. Chem.*, *27*, 493)

Fig. 2-26. The β-barrel structure of triosephosphate isomerase.

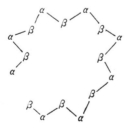

and is regarded as a structure consisting of four connected basic units of $^\beta{\searrow}_\alpha{\nearrow}^\beta{\searrow}_\alpha$.

In α/β proteins the arrangement of α-helices is also parallel.

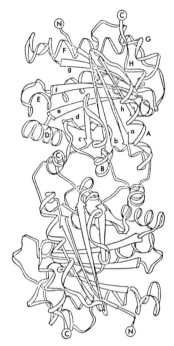

Fig. 2-27. Structure of chicken triosephosphate isomerase. β-strands are represented by a-h and α-helices by A-H. N, amino terminus; C, carboxyl terminus. (Phillips, D.C. *et al.* (1977) *Biochem. Soc. Trans., 5,* 642)

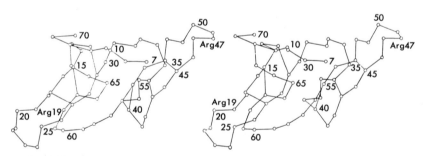

Fig. 2-28. Stereo drawing of the α-carbon backbone of Bowman-Birk trypsin inhibitor. (Suzuki, A. *et al.* (1987) *J. Biochem., 101,* 267)

The α-helix forms a dipole with a partial positive charge at the N-terminal end and a partial negative charge at the C-terminal end due to the aligned dipoles of the peptide units (see Fig. 2-29). When

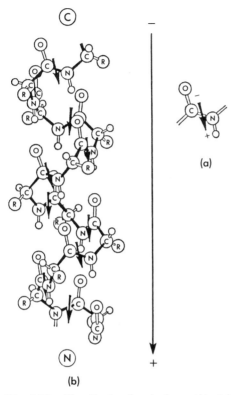

(a)

(b)

Fig. 2-29.   The dipole of a single peptide (a) and the dipole of the α-helix (b). The helix dipole has approximately a 0.5-unit positive charge at the N-terminus and approximately a 0.5-unit negative charge at the C-terminus.

some α-helices are packed together as in the α/β barrel, the large electric field induced by the helix dipoles facilitates binding of the coenzyme molecule, which bears a negative charge, to the N-terminal end of the barrel bearing partial positive charges (Hol, 1985).

e) Coil proteins

There are probably very few proteins which belong to this class. Only Bowman-Birk protease inhibitor has been reported (Fig. 2-28).

The secondary structures of many kinds of proteins have been beautifully depicted by Richardson (1981). This is worth reading in order to understand the structures of various proteins and their classification.

2-3. RELATIONSHIP BETWEEN SECONDARY STRUCTURE AND AMINO ACID SEQUENCE

As described above, different proteins have different secondary structures, and the same applies to different segments of protein polypeptide chains. What factors determine whether some segments have α-helices, some β-structures, and others reverse-turn structures? One of the most decisive factors is the amino acid sequence; certain residues favour or rule out the formation of these structures. The relationship between secondary structure and amino acid sequence has been inferred from examination of the frequency of occurrence of each of the 20 amino acid residues in a particular secondary structure. The frequencies with which amino acid residues occur in the α-helix, β-sheet and reverse turn, obtained by analysis of more than 50 different proteins, are given in Table 2-4.

TABLE 2-4

Capacities of Amino Acid Residues to Form α-Helix, β-Sheet and Reverse Turn

Residue	Helix	β-Sheet	Reverse turn	Residue	Helix	β-Sheet	Reverse turn
Glu	h	b	i	Ile	i	h	b
Ala	h	i	b	Asp	i	b	h
Leu	h	i	b	Thr	b	h	i
His	h	i	b	Ser	b	i	h
Met	h	i	b	Arg	i	i	(i) (b)
Gln	h	b	i	Cys	(h)	b	(b)
Trp	i	(h)	(b)	Asn	i	b	h
Val	i	h	b	Tyr	b	h	i
Phe	i	h	b	Pro	b	b	h
Lys	h	b	i	Gly	b	i	h

h, former; i, indifferent; b, breaker.
Levitt, M. (1978) *Biochemistry, 17*, 4277.

In this table, the β-sheet propensities have been determined without distinction between parallel and antiparallel β-sheets.

As can be seen in Table 2-4, each of the amino acid residues, except for Arg, favours only one of the three possible types of secondary structure, and not the other two. The Arg residue has a low frequency of occurrence in any of the secondary structures. Amino acid residues with side chains branched at C_β (Val, Ile and Thr) and three aromatic residues (Phe, Tyr and Trp) favour the β-sheet. The His residue has a low β-sheet propensity. Amino acid residues with short polar side chains (Ser, Asp and Asn), Gly and Pro, have high frequencies of occurrence in reverse-turns. The other residues favour the α-helix. The short polar side chains of Ser, Asp and Asn form hydrogen bonds with the C=O and N-H groups of the polypeptide main chain, and stabilize the reverse-turn structure. Since Gly has no side chain, it can form the type II reverse-turn. The Pro residue, with its ring structure, fits well into the reverse-turn structure. The Gly residue, with no side chain, is important for packing the polypeptide chains together and allows conformations that are usually prohibited. Examination of the amino acid sequences of cytochromes *c* from 30 different species shows that almost all of the Gly residues are conserved. The Gly residue is thus considered to be structurally important.

Ser, Thr or Tyr, with an OH group in the side chain, Gly with only a hydrogen atom, and Pro with no hydrogen at the nitrogen atom, break the α-helix. The OH group of Ser or Thr can be brought so close to the polypeptide main chain that it forms a hydrogen bond with the C=O group of the peptide, and thus formation of a hydrogen bond between the C=O and N-H groups within the main chain is prevented. The side-chain OH group of Tyr cannot come close to the main chain, but interferes with helix formation. Amino acid residues with a charged group in the side chain (Glu, Asp and Lys) and residues with an amide in the side chain (Gln and Asn) break the β-sheet structure. Pro also has a low propensity for β-sheet formation. The frequency at which non-polar residues occur in the reverse-turn structure is low. One of the reasons for this may be that reverse-turns are usually located on the surface of the protein molecule, where the environment is unfavourable for hydrophobic residues.

Chou and Fasman (1978) attempted to predict secondary structures on the basis of the primary structure and the frequency of occurrence of each of 20 amino acids in the α-helix, β-sheet and reverse-turn. Twenty-nine proteins and 4,741 amino acid residues were used to determine the frequencies. As an example, let us explain how to obtain the frequency of occurrence of Ala in the α-helix. The total number of Ala residues used was 434. Of these, 234 were found to occur in α-helices. Therefore, the frequency of occurrence of Ala in the α-helix (f_α) is $234/434 = 0.539$. The total number of amino acid residues found in α-helices amounts to 1,798. Thus the average frequency of the total amino acid residues found in α-helices ($\langle f_\alpha \rangle$) is $1,798/4,741 = 0.379$. If P_α is defined as $f_\alpha / \langle f_\alpha \rangle$ ($= 0.539/0.379 = 1.42$), the P_α value can be used as a parameter for defining the frequency of Ala occurrence in the α-helix. Since the P_α value for Ala is greater than unity, the Ala residue is found to favour the α-helix. If the P_α value of an amino acid residue is less than unity, it does not favour the α-helix. In a similar way, the frequency parameters of amino acid residues for the β-sheet (P_β) and reverse-turn (P_t) were determined (Table 2-5). Chou and Fasman (1978) classified the twenty amino acid residues into six types (H, h, I, i, b, and B) according to their propensities for α-helix or β-sheet. The residues with symbols H_α (or H_β), h_α (or h_β) or I_α (or I_β) favour the α-helix (or β-sheet) and the helix (or β)-forming capacity decreases in the order $H > h > I$. The residues with symbol b_α (or b_β) or B_α (or B_β) break the α-helix (or β-sheet). The breaking capacity is greater for B than for b. Residues with symbol i have no predilection for formation of an α-helix or β-sheet. The propensities of the amino acid residues for the α-helix, β-sheet and reverse-turn given in Table 2-5 are in general agreement with those given in Table 2-4.

Chou and Fasman (1978) also determined the frequencies of occurrence for amino acid residues in the amino-terminal and carboxyl-terminal regions of the α-helix or β-sheet. They found that Glu and Asp, bearing a negative charge, favour the amino-

TABLE 2-5

Frequency Parameters for Occurrence of Amino Acid Residues in α-Helix, β-Sheet and Reverse-turn

P_α		P_β		f_i	f_{i+1}	f_{i+2}	f_{i+3}	P_t
Flu 1.51	H$_\alpha$	Val 1.70	H$_\beta$	Asn 0.161	Pro 0.301	Asn 0.191	Trp 0.167	Asn 1.56
Met 1.45		Ile 1.60		Cys 0.149	Ser 0.139	Gly 0.190	Gly 0.152	Gly 1.56
Ala 1.42		Tyr 1.47		Asp 0.147	Lys 0.115	Asp 0.179	Cys 0.128	Pro 1.52
Leu 1.21		Phe 1.38	h$_\beta$	His 0.140	Asp 0.110	Ser 0.125	Tyr 0.125	Asp 1.46
Lys 1.16	h$_\alpha$	Trp 1.37		Ser 0.120	Thr 0.108	Cys 0.117	Ser 0.106	Ser 1.43
Phe 1.13		Leu 1.30		Pro 0.102	Arg 0.106	Tyr 0.114	Gln 0.098	Cys 1.19
Gln 1.11		Cys 1.19		Gly 0.102	Gln 0.098	Arg 0.099	Lys 0.095	Tyr 1.14
Trp 1.08		Thr 1.19		Thr 0.086	Gly 0.085	His 0.093	Asn 0.091	Lys 1.01
Ile 1.08		Gln 1.10		Tyr 0.082	Asn 0.083	Glu 0.077	Arg 0.085	Gln 0.98
Val 1.06		Met 1.05		Trp 0.077	Met 0.082	Lys 0.072	Asp 0.081	Thr 0.96
Asp 1.01	I$_\alpha$	Arg 0.93	i$_\beta$	Gln 0.074	Ala 0.076	Thr 0.065	Thr 0.079	Trp 0.96
His 1.00		Asn 0.89		Arg 0.070	Tyr 0.065	Phe 0.065	Leu 0.070	Arg 0.95
Arg 0.98	i$_\alpha$	His 0.87		Met 0.068	Glu 0.060	Trp 0.064	Pro 0.068	His 0.95
Thr 0.83		Ala 0.83		Val 0.062	Cys 0.053	Gln 0.037	Phe 0.065	Glu 0.74
Ser 0.77		Ser 0.75	b$_\beta$	Leu 0.061	Val 0.048	Leu 0.036	Glu 0.064	Ala 0.66
Cys 0.70		Gly 0.75		Ala 0.060	His 0.047	Ala 0.035	Ala 0.058	Met 0.60
Tyr 0.69	b$_\alpha$	Lys 0.74		Phe 0.059	Phe 0.041	Pro 0.034	Ile 0.056	Phe 0.60
Asn 0.67		Pro 0.55	B$_\beta$	Glu 0.056	Ile 0.034	Val 0.028	Met 0.055	Leu 0.59
Pro 0.57	B$_\alpha$	Asp 0.54		Lys 0.055	Leu 0.025	Met 0.014	His 0.054	Val 0.50
Gly 0.57		Glu 0.37		Ile 0.043	Trp 0.013	Ile 0.013	Val 0.053	Ile 0.49

f_i, f_{i+1}, f_{i+2} and f_{i+3} represent the frequency of occurrence of residues at positions i, $i+1$, $i+2$ and $i+3$, respectively, in the reverse-turn. P_t represents the value of P for four residues in the reverse-turn. H, h and I represent the residues which favour the α-helix or β-sheet. The forming capacity decreases in the order H$>$h$>$I. B and b represent the residues which break the α-helix or β-sheet. The breaking capacity is greater for B than for b. i represents the residues which have no predilection for the formation of the α-helix or β-sheet.

Chou, P.Y. & Fasman, G.D. (1978) *Adv. Enzymol.*, **47**, 45.

terminal end of the helix, and that Lys, His and Arg, bearing a positive charge, favour the carboxyl-terminal end. The implications of these results are discussed in Section 2-4. They also found that Pro favours the amino-terminal end of the helix and is not found at all at the carboxyl-terminal end, and that Asn and Trp tend to appear at the amino-terminal end of the helix, whereas Met, Phe, Val and Leu occur at the carboxyl-terminal end. For the β-sheet conformation, Gln, Met and Thr occur at the amino-terminal end more frequently than at the carboxyl-terminal end, and Asn and

His tend to appear at the carboxyl-terminal end. These results are useful for determining where the formation of the α-helix or β-sheet begins and ends.

Using the amino acid sequence and the propensities of the amino acid residues for the α-helix and β-sheet, Chou and Fasman (1978) proposed a method for predicting the secondary structures of proteins. Although many methods for predicting secondary structures from the primary structure have appeared since then, the Chou-Fasman method is simple and has become the most popular. Readers should refer to a book entitled "Prediction of Protein Structure and the Principles of Protein Conformation" edited by G.D. Fasman (1989).

None of the methods proposed so far can completely predict the secondary structure from the primary structure alone. This suggests that the secondary structures of proteins are not determined merely by the amino acid sequence, but also by other factors, such as interactions between distantly separated polypeptide segments. This is discussed with regard to helix formation in the following section.

2-4. HELIX FORMATION

First, we will describe the Zimm-Bragg theory (1959) of the helix-coil transition. The reaction for formation of a helix (H) from a random coil (C) is expressed by the following equation

$$C \underset{k_b}{\overset{\sigma k_f}{\rightleftharpoons}} H_1 \underset{k_b}{\overset{k_f}{\rightleftharpoons}} H_2 \underset{k_b}{\overset{k_f}{\rightleftharpoons}} \text{---} \underset{k_b}{\overset{k_f}{\rightleftharpoons}} H_{n-1} \underset{k_b}{\overset{k_f}{\rightleftharpoons}} H_{n-1}, \qquad (2.1)$$

where k_f is the rate constant for adding one residue to the helix and k_b is the rate constant for removing one residue. The equilibrium constant s is expressed by

$$s = k_f / k_b. \qquad (2.2)$$

Both k_f and k_b have similar values in the range of $10^8-10^{11}/\text{sec}$, depending on the kind of amino acid residue. σ is called the nucleation factor, which is a measure of the difficulty in forming the nucleation of the first helix. The value of σ is within the range 10^{-5}–10^{-2}. Formation of the first helix in the random coil conformation (C) is the slowest and energetically most unfavourable step. The growth of the second and following helix turns is a relatively favourable reaction. Two reasons can be considered for this. (1) As described above, in the α-helix a hydrogen bond is formed between main chain atoms that are four residues apart. Therefore, the nucleus for helix formation is first formed only when residues which form a hydrogen bond, and the three residues between them, are fixed. Addition of one residue to a helix that has already formed occurs more easily. (2) In the α-helix, all the peptide bonds are oriented in the same direction (see Fig. 2-29) and the interactions between the neighbouring parallel peptides are unfavourable when the first single turn of the helix is formed. It is thus difficult for the first helix to form. Once the first one-turn helix is formed, the next residue is easily added, because the dipole interactions occur favourably in a head-to-tail manner.

Because of the difficulty involved in helix nucleation, a polypeptide chain has a tendency to assume either a complete helix (H_n) or a complete random coil conformation (C). The equilibrium constant K_n between H_n and C is expressed by the following equation

$$K_n = \frac{(H_n)}{(C)} = \sigma s^n. \tag{2.3}$$

If σ and s are taken as 10^{-3}–10^{-4} and 1, respectively, the value of n must be large in order for the helix to be stable ($K_n > 1$). For example, when $\sigma = 10^{-4}$, the average length of the polypeptide chain should be greater than 100 residues. Therefore, the Zimm-Bragg theory predicts that a short polypeptide chain consisting of 20 residues or less cannot form a helix.

Figure 2-30 shows the relationship between the helix formation parameter (s in Eq. (2.2)) for various amino acid residues obtained from studies with synthetic polypeptides, and the helix

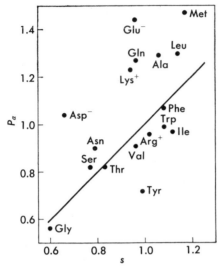

Fig. 2-30. Relation between helix formation parameter s and helix propensity P_α. (Creighton, T.E. (1984) "Protein Structure and Molecular Properties," W.H. Freeman & Co., New York, p. 331)

propensities (P_α in Table 2-5) obtained from proteins. It is clear that these two parameters are not correlated.

However, the helices which exist in proteins consist of about 10 residues on average, and it has often been observed that when a helical segment in a protein is isolated as a fragment by limited proteolysis or chemical treatment, the fragment no longer has the helical conformation but assumes a random coil conformation. A number of factors might contribute to this stabilization of the helix in the context of the native protein, viz. (1) interaction with peptide segments which are far apart in the sequence, (2) the presence of particular amino acid residues at the termini of the helix and (3) interactions such as salt bridge formation between positively charged and negatively charged side chains in the helix.

Limited proteolysis of ribonuclease A with subtilisin yields a fragment called S-peptide, consisting of the residues from the N-terminus to Ala 20 (Fig. 2-31). Cyanogen bromide treatment of ribonuclease A yields a fragment called C-peptide, consisting of the residues from the N-terminus to Met 13 (converted to homoserine

```
 1            5              10
Lys—Glu—Thr—Ala—Ala—Lys—Phe—Glu—Arg—
             15             20
Gln—His—Met—Asp—Ser—Ser—Thr—Ser—Ala—Ala
```

Fig. 2-31. The amino acid sequence of S-peptide. α-Helix is formed from Thr 3 to Met 13 in the ribonuclease A molecule. C-peptide obtained by cyanogen bromide treatment of ribonuclease A is from Lys 1 to Met 13, which is converted to homoserine lactone.

lactone). It was found that these two peptides form a helix at 5°C (Bierzyński *et al.*, 1982; Shoemaker *et al.*, 1985). This finding cannot be explained on the basis of the Zimm-Bragg theory and suggests that some other factors contribute to the stabilization of the helical structures of these short peptides. The stability of a peptide, acetyl-Ala-Glu-Thr-Ala-Ala-Ala-Lys-Phe-Leu-Arg-Ala-His-Ala-CONH$_2$, which is analogous to C-peptide, has been studied at various pH values (Shoemaker *et al.*, 1987). By converting the N-terminal residue to succinyl-Ala, Ala and Lys, by which the charge on this residue is changed from -1 to $+2$, or by converting Glu 2 to Ala, or His 12 to Ala, it was found that the interaction between negative charges on the N-terminus and the helix dipole, and the interaction between positive charges on the C-terminus and the helix dipole are important for stabilization of the helix. Tonan *et al.* (1990) have recently prepared some peptides including the helical segment of chicken pancreatic polypeptide (see Fig. 2-12) and studied their conformations and stability. They found that the interactions of a negative charge at the N-terminal region and a positive charge at the C-terminal region with the helix dipole are important for stabilizing the helix. It is particularly interesting to note that removal of the amide group of the C-terminal residue destabilizes the helix. It has recently been reported that a short peptide corresponding to the α-helical region of bovine pancreatic trypsin inhibitor shows partial folding, and that the helix is stabilized by interactions between the charged groups and the helix dipole (Goodman & Kim, 1989). These observations are in accord with the results obtained by Chou and Fasman (1978), *i.e.,* that the acidic residues and basic residues tend to occur in the N-terminal and C-terminal regions, respectively, of the α-helix (see Section 2-3).

The salt bridge also stabilizes the helical conformation of short peptides. It has been shown that a salt bridge between Glu$^-$ and Lys$^+$ is important for stabilization of the helical conformation of AcAla-Glu-Ala-Ala-Ala-Lys-Glu-Ala-Ala-Ala-Lys-Glu-Ala-Ala-Ala-Lys-Ala-NH$_2$ (Marqusee & Baldwin, 1987), and that the side-chain interaction between Glu 2$^-$ and Arg 10$^+$ and possible interaction between Phe 8 and His 12$^+$ are important for stabilization of the C-peptide of ribonuclease A (Fairman *et al.*, 1990; Shoemaker *et al.*, 1990).

When the stability and formation of an α-helix are affected by the presence of charges at both of its termini, and by a salt bridge, phosphorylation of a side chain, binding or access of an ion, or a change in the membrane potential of the cell membrane might trigger a helix-coil transition in the local region of the protein molecule. Such a local conformational change may lead to a drastic conformational change in the whole molecule, which may, in turn, modify the function of the protein. Sprang *et al.* (1988) reported that activation of glycogen phosphorylase by phosphorylation of Ser 14 is attributable to formation of a helical structure around the phosphorylation site.

Perutz *et al.* (1965) examined the secondary structure of globin in relation to its amino acid sequence. Figure 2-32 shows this secondary structure; α-helical segments are represented as sine waves and non-helical segments as straight lines. The sine waves are drawn so that the top of each wave points to the interior of the globin molecule. It may be seen that in the helical segments non-polar residues occur once every three or four residues, and point to the interior of the molecule. This means that the non-polar residues lie on one side of the helix, since one turn of the α-helix consists of 3.6 residues.

Figure 2-33 shows the helical region of chicken pancreatic polypeptide represented by a wheel. The α-helix has a rotation of 100° per residue, and it can be seen clearly from this figure that hydrophobic residues are located on one side of the helix. Thus,

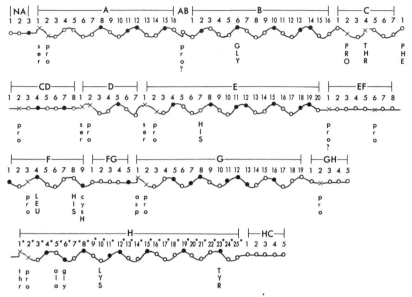

Fig. 2-32. Secondary structure of globin. Sine waves represent α-helical segments and straight lines non-helical segments. Black circles indicate sites where only non-polar residues occur and crosses indicates sites where prolines or combinations of prolines with Ser, Thr, Asp or Asn occur. All other sites are represented by clear circles. Residues in capital letters are invariant. The sine waves are drawn so that the top of each wave points to the inside of the globin chain. (Perutz, M. F. *et al.* (1965) *J. Mol. Biol., 13,* 669)

when the amino acid residues are arranged such that a hydrophobic residue appears once every three or four residues in the sequence, this segment will tend to form such an amphipathic α-helix at the surface of the protein.

2-5.   SECONDARY STRUCTURES AND FUNCTIONS

Proteins with similar primary structures generally have similar secondary and tertiary structures. Myoglobin and each of the subunits of haemoglobin have very similar (α-protein) structures called globin-folds, which share a common function to bind oxygen. The primary structures and conformations of the serine en-

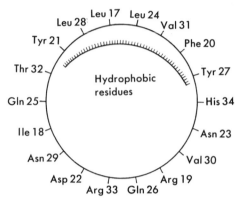

Fig. 2-33. Helical wheel representation of the α-helical segment of chicken pancreatic polypeptide. (Wood, S.P. *et al.* (1977) *Eur. J. Biochem.*, *78*, 119)

zymes, chymotrypsin, trypsin and elastase, are also very similar. Subtilisin, although also a serine enzyme, differs from the above three in its primary, secondary and tertiary structures. However, its active site conformation is very similar to those of the other three and is an example of convergent evolution. The primary, secondary and tertiary structures of horse, tuna and bonito cytochrome c are similar. They are also similar to those of cytochrome c of photosynthetic and nonphotosynthetic bacteria. These common tertiary structures are called cytochrome folds. The similarity between the primary structure of bovine cytochrome c and that of bacterial cytochrome c_{551} is not markedly apparent when they are simply compared pairwise from their N-termini. Nevertheless, the secondary and tertiary structures of these two proteins are very similar. When the primary structures are compared on the basis of the secondary and tertiary structures however, it is found that there are deletions of amino acid residues in some parts of cytochrome c_{551}. When these deletions are taken into account, the similarity between the primary structures of these two proteins becomes apparent. This is an example of two proteins whose primary structures can only be compared meaningfully after elucidation of their secondary and tertiary structures. Thus, primary structure similarity is not always detected easily, although procedures for automatic sequence align-

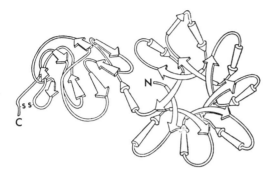

Fig. 2-34.   Structure of Taka-amylase A. (Kindly provided by M. Kusunoki)

ment that take account of insertions and deletions have been developed to improve detection of sequence similarity.

The eight-fold α/β-barrel fold, which was first discovered in triosephosphate isomerase (Fig. 2-27) has also been found in pyruvate kinase, enolase, aldolase, Taka-amylase A, xylose isomerase, glycolate oxidase, ribulose-biphosphate carboxylase (Rubisco), muconate lactonizing enzyme, tryptophan synthase, trimethylamine dehydrogenase, flavocytochrome b_2 and N-(5'-phosphoribosyl) anthranilate isomerase-indole-3-glycerol-phosphate synthase (Lebioda & Stec, 1988). It is interesting to note that most of the glycolytic enzymes have the α/β-barrel fold. The barrel shapes of proteins of this class vary. For instance, the cross-sectional diameter of the barrel of glycolate oxidase is 14.5 Å, whereas that of triosephosphate isomerase is 16.5×11.5 Å. The Taka-amylase A molecule consists of two domains (Matsuura et al., 1984), one with an α/β-barrel fold and the other showing a conformation very similar to each of the domains of the immunoglobulin molecule (Fig. 2-34). Primary sequence similarities scarcely exist between the proteins of this class.

The structures of various aminoacyl-tRNA synthetases are very interesting. Dimeric tyrosyl-tRNA synthetase from *Bacillus stearothermophilus*, the monomeric tryptic fragment of *Escherichia coli* methionyl-t-RNA synthetase and glutamyl-t-RNA synthetase all have a common β-α-β structural motif, although they have no marked sequence similarity. In contrast, the structure of seryl-tRNA from *Escherichia coli* is quite different from the above three synthetases, and lacks the β-α-β motif (Cusack *et al.*, 1990). It can be easily understood that proteins with similar primary structures will assume very similar secondary and tertiary structures. However, there are many instances in which proteins with no apparent sequence similarity have very similar secondary and tertiary structures. For these proteins, it is difficult to determine whether the evolutionary processes have been divergent or convergent. Generally, in the processes of protein evolution, secondary and tertiary structures are conserved to a greater extent than primary structure.

Proteinase B of *Streptomyces griseus*, with 186 amino acid residues, is smaller than chymotrypsin, which consists of 234 resi-

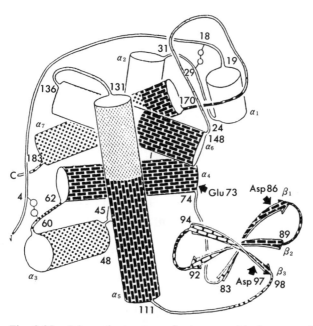

Fig. 2-35. Schematic structure of goose egg-white lysozyme. Bricks, the segments common to goose, hen and T4 lysozymes; dots, the segments common to goose and hen lysozymes; dashes, the segments common to goose and T4 lysozymes. Unshaded parts represent the segments of goose lysozyme only. (Weaver, L.H. *et al.* (1985) *J. Mol. Evolut.*, *21*, 97)

dues. Deletions and insertions of amino acid residues occur only at the surface of these molecules, and the secondary and tertiary structures and the active site conformations are very similar. However, only 13% of the residues are identical, even when the primary structures are compared on the basis of secondary and tertiary structure.

It is also very interesting to compare the secondary and tertiary structures of hen egg-white, goose egg-white and T4 phage lyso-zymes. Chicken-type lysozymes all consist of 129 amino acid resi-dues. Goose-type lysozymes consist of 180 residues. The egg-white of the black swan (*Cygnus altretus*) contains both chicken-type lysozyme and goose-type lysozyme. The primary structure of hu-man lysozyme (130 residues) is similar to that of chicken-type lysozyme, and differs at only 52 positions. T4 phage lysozyme consists of 180 residues. Figure 2-35 shows the polypeptide chain of goose egg-white lysozyme and the segments in which the secondary structure is common to goose, hen and T4 lysozymes, and those in which the secondary structure is common to any two of the three lysozymes (Weaver *et al.*, 1985). There are 90 and 91 α-carbon atoms that correspond structurally between goose and hen lyso-zymes and between goose and phage lysozymes, respectively. However, there is no sequence similarity in these regions. The catalytic groups of hen egg-white lysozyme are Glu 35 and Asp 52. At the corresponding positions in the three-dimensional structures, Glu 11 and Asp 20 are present in phage lysozyme, and Glu 73 and Asp 86 in goose egg-white lysozyme. Since the positional relation-ship of these groups is very similar in the three lysozymes, Glu 11 and Asp 20 for phage lysozyme and Glu 73 and Asp 86 for goose lysozyme are considered to be the catalytic groups. Although no sequence similarity exists, similarities in secondary and tertiary structure, catalytic groups and active site conformation suggest that these three lysozymes have evolved divergently from a common ancestral protein. Lysozyme obtained from *Streptomyces ery-thraeus* consists of 185 residues and shows no similarity in second-ary or tertiary structure to any of the above three lysozymes.

α-Lactalbumin from baboon milk consists of 123 amino acid residues, of which 44 are identical with hen egg-white lysozyme. The overall structure of α-lactalbumin is very similar to that of hen egg-white lysozyme (Acharya *et al.*, 1989), but the functions of the two enzymes are quite different.

Mandelate racemase and muconate lactonizing enzyme catalyze quite different reactions. Mandelate racemase has no muconate lactonizing activity, and muconate lactonizing enzyme has no mandelate racemase activity. Nevertheless, both enzymes possess very similar secondary, tertiary and quaternary structures. In these enzymes, 26% of the residues are identical in sequence (Neidhart *et al.*, 1990).

Recently it has been found that β-lactoglobulin, serum retinol binding protein, biliverdin binding protein and rat intestinal fatty acid binding protein have very similar structures. Eight or ten antiparallel β-strands form two β-sheets. The folding of β-lactoglobulin is shown in Fig. 2-36. The overall shape of the structure resembles a clam shell and has been called a "β-clam" (Sacchettini *et al.*, 1988). These proteins have no hydrophobic core, and a large solvent-filled internal cavity is formed instead. The

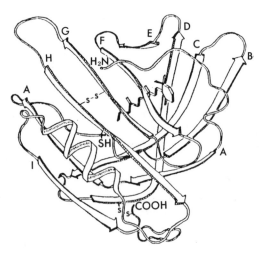

Fig. 2-36. Secondary structure of β-lactoglobulin. The binding site of retinol is also shown. (Papiz, M.Z. *et al.* (1986) *Nature, 324*, 383)

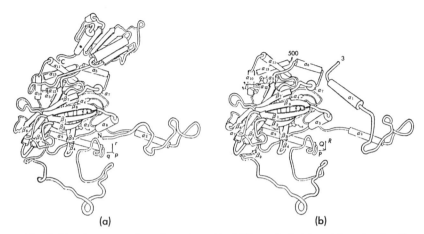

Fig. 2-37. Structure of catalases from *Penicillium vitale* (a) and from bovine liver (b). (Melik-Adamyan, W.R. *et al.* (1988) *J. Mol. Biol., 188*, 63)

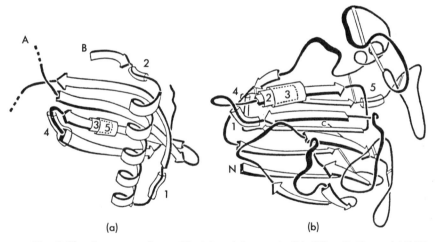

Fig. 2-38. Structures of monellin (a) and thaumatin (b). (Kim, S.-H. *et al.* (1988) *TIBS, 13-January,* 13)

solvent in the cavity is displaced by ligand binding, and the β-clam structure seems to be important for binding and retaining hydrophobic ligands inside the molecule. The sequence similarities between the β-clam proteins are very low.

Cytochrome b_{562}, cytochrome c', haemerythrin and the coat protein of tobacco mosaic virus all have $\alpha\alpha'$-supersecondary structures, but their functions are quite different. Each of the immunoglobulin domains and a subunit of superoxide dismutase have similar secondary structures but have no common function. Wheatgerm agglutinin, snake venom protein, pollen allergen, hevein and neurophysin all have very similar secondary structures.

Catalase from bovine liver has a structure similar to that from *Penicillium vitale*, but the fungal catalase has an additional structure similar to that of flavodoxin (Fig. 2-37).

Monellin and thaumatin are proteins obtained from the African berry. They have a very high affinity for the sweet receptor and are about 100,000 times sweeter than sugar on a molar basis and about 4,000 times sweeter on a weight basis. Although both proteins are very sweet, there are neither significant sequence similarities nor secondary or tertiary structural similarities between them (Fig. 2-38). However, antibodies against thaumatin compete for monellin and *vice versa*. This is the first known example of proteins that have immunological cross-reactivity without sequence or structural similarities.

We have described examples of proteins that have similar secondary and tertiary structures. Functional similarities do not always mean structural similarities, and likewise the opposite does not necessarily apply. Structural similarities between proteins are of interest in connection with questions of protein evolution and stability.

2-6. METHODS OF STUDYING SECONDARY STRUCTURES

X-ray crystallography is the most powerful method for determining which segment of the polypeptide chain of a protein molecule has an α-helical, β-sheet or reverse-turn structure. However, this method cannot always be employed, because many proteins are difficult to crystallize. Recently, application of the NMR method to the determination of the secondary and tertiary structures of proteins in solution has advanced considerably (Wüthrich, 1986).

Circular dichroism (CD) spectroscopy is frequently used to determine the secondary structural components of proteins in solution. Figure 2-39 shows the CD spectra of the four basic structures of polypeptide chains. The CD spectrum of the α-helix shows a negative maximum at 222 nm due to the n-π^* transition of the peptide group and another at 208 nm due to the π-π^* transition, and also a positive maximum at 190 nm due to the π-π^* transition. The CD spectrum of the β-sheet shows a negative maximum at 218 nm due to the n-π^* transition and a positive maximum at 195 nm due to the π-π^* transition. The CD spectrum of the reverse turn differs according to the model peptide used.

It can be determined readily whether a protein has an abundance of α-helix or an abundance of β-sheet. The CD spectra of α-proteins, β-proteins, $\alpha + \beta$ proteins and α/β proteins described in Section 2-2 are shown in Fig. 2-40. However, it has not been clearly established whether the secondary structural classification of proteins can be made on the basis of CD spectra alone.

It is also difficult to estimate the content of α-helix or β-sheet in a protein on the basis of CD data alone. Previously, the CD spectra of α-helix, β-sheet and the random coil of synthetic poly-

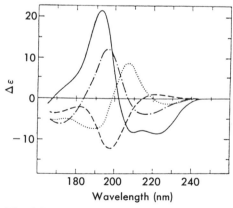

Fig. 2-39. The CD spectra of α-helix (——), β-sheet (—·—), reverse turn (·····) and random coil (---). (Johnson, W.C. Jr. (1988) *Annu. Rev. Biophys. Chem., 17*, 145)

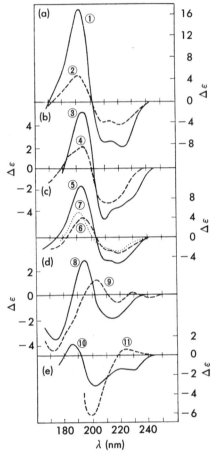

Fig. 2-40. The CD spectra of various proteins. (a) α-proteins. ① myoglobin; ② cytochrome c. (b) $\alpha + \beta$ proteins. ③ hen egg-white lysozyme; ④ ribonuclease A. (c) α/β proteins. ⑤ triosephosphate isomerase; ⑥ flavodoxin; ⑦ subtilisin BPN'. (d) β-proteins. ⑧ prealbumin; ⑨ Bence-Jones protein. (e) β-proteins. ⑩ α-chymotrypsin; ⑪ soybean trypsin inhibitor. (Manavalan, P. & Johnson, C. Jr. (1983) *Nature, 305,* 831)

peptides such as poly-L-glutamic acid and poly-L-lysine were used as reference standards for estimating the content of each type of protein secondary structure. Recently the CD spectra of pure α-helix, β-sheet and random coil constructed using CD spectra of proteins with known secondary structures have been used as reference standards (Chang *et al.*, 1978). Other methods for estimating secondary structures using CD spectra are referred to in the articles by Provencher and Glöckner (1981), Manavalan and Johnson (1983), and Johnson (1988).

None of these methods provides a precise determination of the secondary structural content of a protein, but even if the content is estimated well, it is impossible to know which segment in a polypeptide chain is α-helix or β-sheet. However, CD spectroscopy is very useful for studying the conformational changes in proteins caused by denaturation or interaction with ligands.

3

Tertiary Structure of Proteins

In Chapter 2, we described the properties of the basic secondary structural elements such as α-helix, β-sheet and reverse turn. A polypeptide chain with regular secondary structure is further folded into a compact globular shape. Tertiary structure refers to the spatial arrangement of a polypeptide chain with its secondary structure. In this chapter, we describe the properties of the tertiary structure of globular proteins. When a ball of string is extended by pulling it from the two opposite ends, the string usually tangles and knots are formed. However, when the polypeptide chain of a globular protein is extended by pulling it from the two termini, the chain does not tangle and is easily unravelled; no knots are formed. This suggests a mechanism by which the polypeptide chain is folded into a compact globular molecule.

3-1. DOMAINS

The folded conformation of proteins with large molecular weights is usually divided into smaller globular folded units (domains). A single domain usually consists of 100–150 amino acid residues and is about 25 Å in diameter. As described in Section 2-2, the light chain of IgG1 immunoglobulin consists of two domains (V_L and C_L) and the heavy chain consists of four domains (V_H, C_H1, C_H2 and C_H3). Each of these domains consists of about 120 amino acid residues. As in the glycolytic enzymes, some enzymes consist of a catalytic domain and a coenzyme-binding domain. The functional domain is sometimes divided into smaller structural domains. The active site of the enzyme is very often located at the site where two or more domains are in contact.

A domain can sometimes be isolated as a fragment by limited proteolysis under appropriate conditions. Each fragment has the same conformation as that existing in the native protein molecule. It is also stable and can be refolded spontaneously from the unfolded state under native conditions.

In the structure of eukaryotic genes, intervening sequences (introns) interrupt the coding regions (exons). It would be very interesting to know why protein-coding eukaryotic genes have this mosaic structure. The four domains of the heavy chain of the immunoglobulin molecule are encoded by four separate exons. However, in haemoglobin and hen egg-white lysozyme, such clear correspondence between exons and domains is not observed. Recently, Go (see Review, 1985) found that in the α and β chains of haemoglobin and hen egg-white lysozyme, there are compact structural units closely related to the exon structure of the genes. Such a structural unit has been called a module by Go (1985), and is defined as a compact substructure of a globular domain. The structural unit of the module is smaller than that of the domain, and consists of 20–40 amino acid residues. For instance, the positions of the introns which interrupt the coding regions of hen egg-white lysozyme correspond to the boundaries between five module structures, M1 (residues 1–30), M2 (residues 31–55), M3 (residues 56–84), M4 (residues 85–108) and M5 (residues 109–129). Each of these modules seems to correspond to a different aspect of the enzyme's function. For instance, the catalytic groups (Asp 52 and Glu 35) exist in module M2 and the binding subsites of the substrate are in modules M2, M3, M4 and M5. Three of the four disulphide bonds seem to serve as linkages between the modules. There are three exons in the genes of the β chain of haemoglobin and myoglobin. The peptides corresponding to these exons are residues 1–30, 31–105 and 106–152, and peptide 31–105 corresponds to the segment which binds the haem group. The other exons mediate $\alpha\beta$, and $(\alpha_1\beta_1)(\alpha_2\beta_2)$ interactions, respectively. Hence the theory for evolution of haemoglobin from myoglobin. Module structures have since been found in a number of other proteins (see Go, 1985; Go & Nosaka, 1987). It has also been shown that the residues at inter-module junctions have low relative accessibilities (see Section 3-2).

Limited proteolysis of horse apomyoglobin with clostripain, which specifically hydrolyzes peptide bonds to arginine residues, yields a peptide fragment from residues 32 to 139. Although this peptide fragment is longer than the module from residues 31 to 105 to which haem binds, its various properties have been studied as a model of the module (De Sanctis *et al.*, 1986, 1988). It was found that this peptide fragment shows a CD spectrum characteristic of an α-helix and binds strongly to protohaem in a 1 : 1 ratio. Since this peptide fragment has properties similar to the original apomyoglobin, it is called mini-globin. Interactions with O_2, CO and a hydrophobic probe ANS (1-anilino-8-naphthalene sulphonate) were found to be very similar to those of myoglobin. These findings suggest that the peptide encoded by the exon is a viable structural and functional unit. One side of a module consists mainly of hydrophobic residues and the other mainly of hydrophilic residues, and the domain is built so that the hydrophobic surface is directed toward the interior, and the hydrophilic surface is exposed to the solvent. It would be interesting to study the properties of other modules isolated as fragments.

3-2. SURFACE AND INTERNAL STRUCTURE OF THE PROTEIN
MOLECULE

As described in Chapter 1, the amino acid residues that
constitute the protein molecule are classified into hydrophobic,
hydrophilic and ionizable residues. Proteins whose structures have
been established by X-ray crystallographic analysis are found to
have hydrophilic and ionizable residues located mainly on the
surface of the molecule and hydrophobic residues mainly buried in
the interior. This distribution of hydrophilic and hydrophobic
residues within the protein molecule was predicted as early as the
1920s by I. Langmuir in the United States and E.K. Rideal in
Britain, prominent pioneers of surface chemistry. They described
the protein molecule as an "oil drop with a polar coat." This
distribution is also called the "nonpolar-in, polar-out" rule. In this
section, the surface and internal structures of the protein molecule
are discussed on the basis of the X-ray crystallographic data.

The accessible surface area is defined as the area described by
the centre of a spherical water molecule of radius R_w (generally
taken as 1.4 Å), which rolls over the van der Waals envelope of the
protein molecule (Fig. 3-1). Table 3-1 shows the accessible surface
area of each of the twenty amino acid residues (X) determined for
a tripeptide, Gly-X-Gly. This value represents the accessible surface
area of a residue exposed completely to solvent water.

The accessible surface area (A_s) in Å^2 of a small monomer
protein can be estimated from its molecular weight M using the
following equation (Miller *et al.*, 1987a)

$$A_s = 6.3 \ M^{0.73}. \tag{3.1}$$

This relation was obtained using 46 monomer proteins with
molecular weights ranging from 4,000 to 35,000.

The total accessible surface area (A_t) in Å^2 of an unfolded
protein is related with its molecular weight by the equation:

$$A_t = 1.48 \ M + 21. \tag{3.2}$$

In this equation, A_s does not become zero when $M = 0$. This can be

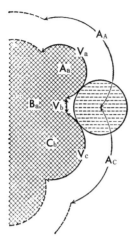

Fig. 3-1. Two-dimensional illustration of accessible surface area. Atoms A and C have accessible surface areas A_A and A_C, respectively. Atom B is sterically obstructed by atoms A and C and has no accessible surface area. (Chothia, C. (1975) *Nature, 254*, 304)

TABLE 3-1
Accessible Surface Area ($Å^2$) of Amino Acid Residues

Residue	Total accessible surface area	Accessible surface area of main chain	Accessible surface area of side chain
Ala	113	40	67
Arg	241	45	196
Asn	158	45	113
Asp	151	45	106
Cys	140	36	104
Gln	189	45	144
Glu	183	45	138
Gly	85	85	
His	194	43	151
Ile	182	42	140
Leu	180	43	137
Lys	211	44	167
Met	204	44	160
Phe	218	43	175
Pro	143	38	105
Ser	122	42	80
Thr	146	44	102
Trp	259	42	217
Tyr	229	42	187
Val	160	43	117

Miller, S. *et al*. (1987) *J. Mol. Biol., 196*, 641.

grasped more easily if we consider that the A_s value occupied by a single point is not zero but has a value of A_s of $4\pi R_w^2$ (25 Å^2), which is close to 21 Å^2. The increase in the accessible surface area with increasing molecular weight is smaller for native proteins than for unfolded proteins. A knowledge of the decrease in the accessible surface area $(A_t - A_s)$ on folding from the completely unfolded state is useful when considering the stability of the protein molecule (Chapter 5).

The relative accessibility of a residue is defined as the ratio of the accessible surface area of the residue in a native protein to that of the residue in the completely unfolded protein. Assuming that residues with a relative accessibility below 5% are buried in the interior of the protein molecule and residues with a relative accessibility above 5% are located on the surface of the molecule, the distribution of each of the amino acid residues between the interior and surface of the molecule has been estimated (Miller *et al.*, 1987a). The results are given in Table 3-2. The partition coefficient (f) of a particular amino acid residue between the interior and the surface of the protein molecule is given by the equation

$$f = \frac{N_s/\Sigma N_s}{N_b/\Sigma N_b},$$ (3.3)

where ΣN_b is the total number of residues buried in the interior of the molecule, N_b is the number of residues of a particular type that are buried, ΣN_s is the total number of residues exposed on the surface of the molecule, and N_s is the number of the same residue type that are exposed. The free energy change of transfer (ΔG_t) of the amino acid residue from the interior to the surface may be estimated using the equation

$$\Delta G_t = -RT \ln f.$$

The ΔG_t values thus obtained are also shown in Table 3-2.

It is found from Table 3-2 that Val, Leu, Ile and Phe occupy about 44% of the internal residues and only about 14% of the surface residues. Cys also prefers the interior. Charged residues, Asp, Glu, Lys and Arg, occupy about 27% of the surface residues

TABLE 3-2

Amino Acid Compositions in the Interior and on the Surface of the Protein Molecule and Free Energy Change of Transfer of Amino Acid Residues from Interior to Surface

Residue	Total number[a]	Interior[a]	Surface[a]	Transfer free energy (kcal/mol)
Ala	8.7	11.0	7.9	0.20
Arg	3.1	0.4	4.0	−1.34
Asn	5.2	2.0	6.3	−0.69
Asp	6.1	2.2	7.4	−0.72
Cys	2.7	5.4	1.8	0.67
Gln	3.6	1.3	4.5	−0.74
Glu	4.9	1.0	6.2	−1.09
Gly	9.0	9.7	8.8	0.06
His	2.2	2.4	2.2	0.04
Ile	4.9	10.5	3.0	0.74
Leu	6.5	12.8	4.3	0.65
Lys	6.7	0.3	8.9	−2.00
Met	1.5	3.0	0.9	0.71
Phe	3.8	7.7	2.5	0.67
Pro	4.0	2.2	4.7	−0.44
Ser	7.0	5.0	8.9	−0.34
Thr	6.4	4.6	7.1	−0.26
Trp	1.6	2.7	1.3	0.45
Tyr	4.4	3.3	4.8	−0.22
Val	6.6	12.7	4.6	0.61
Total	5,436	1,396	4,040	
N-terminal				−1.25
C-terminal				(−2.0)

[a] The compositions are expressed as percentages. The residues with less than 5% relative accessibility are assumed to be buried in the interior and those with more than 5% relative accessibility are assumed to be located on the surface of proteins. Thirty-seven proteins were included in the study.

Miller, S. *et al*. (1987) *J. Mol. Biol., 196*, 641.

and only about 4% of the internal residues. Ala, Gly, Ser and Thr are distributed equally between the surface and interior.

The ΔG_t values for the N- and C-terminal residues are large negative values. This indicates that the N- and C-terminal residues of most proteins are on the surface of the molecule and thus exposed to solvent. Exceptionally, the N-terminal residues of serine proteases and phospholipase A_2 are buried in the interior of the molecule, but the C-terminal residues of all proteins are exposed to solvent. It is worth noting that 76% of all N-terminal amino acids are Ala, Ile, Lys and Met, and that Asn frequently occurs at C-termini.

In Chapter 1 we described the hydrophobic scales of the side chains of amino acid residues (Δg_t) determined from the partition coefficients between octanol and water (Table 1-4). In Fig. 3-2, these Δg_t values are plotted against the ΔG_t values determined from the distribution of the amino acid residues between the interior and surface of the molecule (Table 3-2). A good correlation between the two values is obtained for hydrophobic residues. Tyr and Pro deviate greatly from the line and occur more frequently on the surface of the protein molecule than would be expected from the Δg_t values. The ΔG_t values for Cys, Val, Ile, Phe, Met and Trp are almost the same (Table 3-2), although the Δg_t values for these residues range from 1.5 to 3.1 kcal/mol.

The accessible surface area of an oligomeric protein is related to its molecular weight by the equation (Miller *et al.,* 1987b),

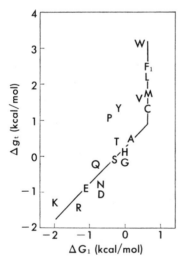

Fig. 3-2. Plot of the hydrophobicities (Δg_t) of amino acid residues (Table 1-4) against the free energy changes (ΔG_t) of transfer of amino acid residues from the interior to the surface of the protein molecule (Table 3-2). (Miller, S. *et al.* (1987) *J. Mol. Biol. 196*, 641)

$$A_s = 5.3 \ M^{0.76}. \tag{3.4}$$

The shapes of oligomer proteins differ, but the same relation (3.4) appears to hold for them all.

If the molecular weight of a subunit is expressed as m, the accessible surface area (a_s) of the subunit of a dimer protein is

$$a_s = \frac{A_s}{2} = \frac{5.3}{2}(2 \ m)^{0.76} = 4.5 \ m^{0.76}. \tag{3.5}$$

The accessible surface area of the subunit of a tetramer protein is

$$a_s = \frac{A_s}{4} = \frac{5.3}{4}(4m)^{0.76} = 3.8 m^{0.76}. \tag{3.6}$$

The accessible surface area of the subunit is thus smaller for a tetramer than for a dimer for any given molecular weight of the subunit.

When the native oligomeric protein molecule is formed from its unfolded molecule, the decrease in the accessible surface area is expressed by

$$A_t - A_s = 1.48 \ M - 5.3 \ M^{0.76}. \tag{3.7}$$

This decrease is the sum of the surface area buried in the interior of the subunits and the surface area buried in the interface between the subunits, the latter ranging from 3% to 40% of the total accessible surface area ($A_t - A_s$) depending on the protein. Therefore, the ratio of the surface area buried in the interior of the subunits to that buried between the subunits differs greatly, but the total accessible surface area buried is the same for any given molecular weight.

Next, we will describe the interior structure of the protein molecule. The packing density of a molecule is defined as the ratio of the volume enclosed by its van der Waals envelope to the volume actually occupied in a crystal or liquid. The packing density of the interior of the lysozyme and ribonuclease A molecules has been found to be 0.75 (Richards, 1974). This means that 75% of the total volume of the interior of the protein molecule is occupied by protein atoms. This packing density is very close to those (0.70–

0.78) of organic crystals and also close to the packing density (0.74) of closely packed spheres of the same size. The packing densities of liquid water and cyclohexane are 0.58 and 0.44, respectively. These facts indicate that the atoms in the interior of the protein molecule are packed as closely as in crystals.

The mean volume occupied by residues in the interior of nine proteins was calculated assuming that atoms with an accessible surface area of less than 5% are buried (Chothia, 1975, 1984). The results are shown in Table 3-3 together with the volumes occupied by amino acids in crystals. The table clearly shows that the mean volume of a residue occupying the interior of a protein is the same

TABLE 3-3
Mean Volumes of Amino Acid Residues Buried in the Interior of Proteins

Residue	Total number of residues	Number of residues with less than 5% accessibility	Mean volume of residue buried ($Å^3$)	Volume of residue in crystal ($Å^3$)[a]
Val	163	91	141.7	143.4
Ala	183	71	91.5	96.6
Ile	106	69	168.8	169.7
Gly	160	60	66.4	66.5
Leu	138	57	167.9	—
Ser	190	46	99.1	102.2
Thr	128	32	122.1	124.3
Phe	60	29	203.4	—
Asp	117	17	124.5	122.0
Cys	34	16	105.6	108.7
Pro	67	16	129.3	124.4
Met	28	14	170.8	176.1
Tyr	98	13	203.6	201.7
Glu	65	13	155.1	143.9
Asn	116	12	135.2	—
Trp	39	9	237.6	—
His	43	8	167.3	166.3
Lys	119	5	171.3	—
Gln	80	5	161.1	148.0
Cys	10	4	117.7	123.1
Arg	63	0	—	—

[a] The volume obtained by subtracting the volume lost by an amino acid on forming a residue (11.1 $Å^3$) from the volume of the amino acid in crystal form.
Chothia, C. (1975) *Nature, 254,* 304.

as that in crystals of the pure amino acid. This suggests that a polypeptide chain that adopts secondary structure must be folded strictly into a crystal-like tertiary structure.* Thus the protein molecule must be depicted more appropriately as a crystal rather than an oil drop. The fact that the interior of the protein molecule is packed as closely as a crystal may be important for the prediction of protein tertiary structure.

The packing density of proteins varies from 0.6 to 0.85 depending on the part of the molecule. Parts where the packing density is lower have greater flexibility and seem to be involved in enzymatic function by controlling the mobility of the active site during catalysis. The packing density of the active site of ribonuclease S is lower than that of its surroundings. In the hen egg-white lysozyme molecule, there is a hydrophobic box around Met 105 formed by the side chains of Trp 23, Trp 28, Trp 111 and Trp 108, and this region has high packing density. On the other hand, there is a cavity in the region surrounded by the side chains of Leu 8, Met 12, Ala 32, Ile 55, Leu 56, Ile 88 and Val 92, which has a low packing density. A site where the packing density is low sometimes has a cavity occupied by a water molecule. This has been observed for chymotrypsin, carboxypeptidase A, trypsin and bovine pancreatic trypsin inhibitor.

3-3. DISULPHIDE BOND

Although the cross-linkages formed by disulphide bonds are an aspect of primary structure, the role of these bonds in protein conformation is discussed here (Thornton, 1981).

The maximum number of disulphide bonds found in a protein is seventeen, and the average number is three. Proteins with more than seven disulphide bonds are rare. Proteins with no disulphide bonds also exist. Generally, extracellular proteins have a greater

---

*This does not imply crystalline regularity—what is interesting is how the protein attains the same packing density as a crystal without the regularity.

number of disulphide bonds than intracellular ones. The intracellular environment is reductive due to the presence of reduced glutathione, and thus the sulphhydryl groups of cysteine residues in proteins remain stable. In the extracelluar environment, however, the presence of oxygen oxidizes sulphhydryl groups to disulphide bonds.

The disulphide bond is usually formed between two cysteine residues located relatively close to each other in the sequence, and most bonds occur between cysteine residues separated by 10–14 residues. This is not always the case, although disulphide bonds formed between cysteine residues more than 150 residues apart are rarely observed. However, there are some instances in which a disulphide bond is formed between a cysteine residue near the N-terminus and one near the C-terminus. One example is hen egg-white lysozyme, which consists of 129 amino acid residues and has four disulphide bonds, one of them between Cys 6 and Cys 127. One of the three disulphide bonds of bovine pancreatic trypsin inhibitor, which consists of 58 residues, is formed between Cys 5 and Cys 58. A disulphide bond of proinsulin, with 85 residues, is formed between Cys 19 and Cys 85, and a disulphide bond of phospholipase A_2, with 123 residues, is formed between Cys 27 and Cys 123. A disulphide bond cannot be formed between two adjacent cysteine residues; this is because the distance between the two adjacent C_α atoms (3.4 Å) is shorter than the distance of the S-S bond. However, a disulphide bond can be formed between two cysteine residues two residues apart. No disulphide bonds have been found to be formed between domains within the same polypeptide chain.

Most disulphide bonds link segments of polypeptide chain that have no secondary structure, and links between two α-helices or two β-strands are rare. Most disulphide bonds are buried in the interior of the protein molecule.

As will be described in Section 5-5, disulphide bonds stabilize the protein molecule by decreasing the conformational entropy in the denatured state. When only insufficient stabilizing interactions due to hydrogen bonds and hydrophobic bonds are available, as in

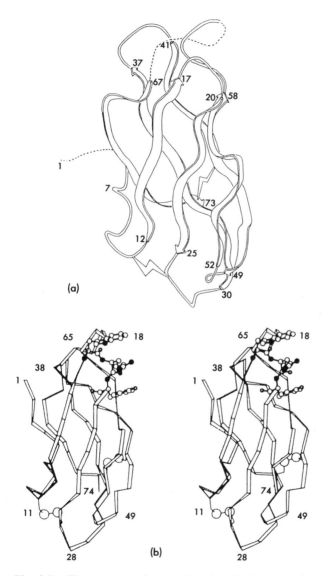

Fig. 3-3. The structure of tendamistat in solution determined by NMR method (a) (Kline *et al*.(1986) *J. Mol. Biol., 189*, 377) and the crystal structure of the inhibitor (b) (Pflugrath, J. *et al.* (1986) *J. Mol. Biol., 189*, 383)

the case of very small proteins, the disulphide bond seems to serve to stabilize the molecule. Small proteins such as erabutoxin, cobra venom toxin, agglutinin, phospholipase A_2, insulin, bovine pancreatic trypsin inhibitor and Bowman-Birk trypsin inhibitor have many disulphide bonds. In the case of ferredoxin and rubredoxin, metal ions instead of disulphide bonds seem to serve to stabilize the molecule.

Chapters 2 and 3 have described the secondary and tertiary structures of proteins obtained by X-ray crystallographic analysis. Recent advances have been made in the determination of protein structures using the NMR method (Wüthrich, 1986). Since protein functions are studied in solution, the NMR method is a very powerful tool for clarifying the relationship between structure and function.*

There are many examples of structures determined both by NMR and X-ray. Figure 3-3 compares the structures of α-amylase inhibitor tendamistat determined by NMR (Kline *et al.*, 1986) and by X-ray (Pflugrath *et al.*, 1986). It can be seen that the structures determined by the two methods are essentially the same, although most surface side chains appears more disordered in solution than in the crystal. The structure of the complement C3a determined by NMR (Nettesheim *et al.*, 1988) is essentially the same as that determined by X-ray (Huber *et al.*, 1980), but the C-terminal region is more ordered in the crystal than in solution, and more structure is seen at the N-terminal region by NMR than X-ray. Comparisons have also been reported for the interleukin-1 and -8 (Clore & Gronenborn, 1991a, b). Most comparisons have indicated that main chains are identical between crystal and solution and that only surface side chains may differ.

---

*Recently it has become possible to study catalyzed reactions in crystals of glycogen phosphorylase b through fast crystallographic data-collection methods (Hajdu *et al.*, 1987).

4

Factors that Stabilize the Protein Molecule

As described in Chapters 2 and 3, all proteins have their own unique secondary and tertiary structures. How are such regular conformations preserved in aqueous solution? If we assume the two-state transition of native (N) $\rightleftharpoons$ unfolded protein (D) (see Section 5-2), the free energy change (ΔG_D) of unfolding may be expressed by

$$\Delta G_D = \Delta H_D - T\Delta S_D. \tag{4.1}$$

Because they are long polymers, proteins can assume a large number of conformations. If the entropy term (ΔS_D) is predominant, the reaction N $\rightleftharpoons$ D proceeds to form the unfolded molecule. However, each protein assumes a unique, folded conformation. This indicates that a positive enthalpy term (ΔH_D) which exceeds the entropy term ($T\Delta S_D$) makes the free energy term (ΔG_D) positive, and the reaction N $\rightleftharpoons$ D proceeds to form the native structure. Because proteins have complex compositions, it seems certain that a number of non-covalent interactions are involved in molecular stabilization. Among these non-covalent interactions, hydrogen bonds, hydrophobic interactions and electrostatic interactions are

considered to be particularly important. Even for these three interactions, however, it is still difficult to appreciate how they are
involved in the stabilization of the protein molecule. Moreover,
when considering protein stability, particular attention should be
given to the effect of solvent water on these interactions.

As will be described in Chapter 5, the free energy change of
unfolding of a protein for the reaction $N \rightleftharpoons D$ in water can be
estimated by analysing the unfolding curves in the presence of
guanidine hydrochloride, urea or heat. It has been found that the
free energy changes for unfolding of proteins in water are very
small, only 5–20 kcal/mol (Table 5-8).* This low free energy of
stabilization of the protein molecule results from compensation of
the various non-covalent interaction energies that stabilize the
native conformation, by the entropy term that stabilizes the unfolded conformation. Thus most globular proteins are marginally
stable relative to their unfolded conformations by only a small
difference in free energy.

4-1. HYDROGEN BOND

Hydrogen bonds are formed between C=O and N–H groups of
the polypeptide main chain in the α-helix and β-sheet. They are
also formed between C=O or N–H groups of the main chain and the
OH groups of the side chains of Ser and Thr or the $CONH_2$ groups
of the side chains of Asn and Gln, and between the functional
groups of side chains. X-ray crystallographic studies show that
almost all the functional groups capable of forming hydrogen
bonds are hydrogen-bonded in the interior of the protein molecule.
It can be appreciated that these many hydrogen bonds would
contribute to the stability of the protein molecule.

In order to understand the properties of the hydrogen bonding
between C=O and N–H groups, the intermolecular association of

---

*The free energy of stabilization of the protein molecule is comparable to that of the
transfer of two or three hydrophobic residues from the interior hydrophobic region to
solvent water (see Table 1-4).

acetamide derivatives through hydrogen bonds has been studied. Figure 4-1 shows the degree of association of N-methylacetamide plotted against its concentration in water, dioxane and carbon tetrachloride (Klotz & Franzen, 1962). As can be seen, the lower the polarity of the solvent, the greater the tendency to associate. In water, extremely high concentrations are required to associate N-methylacetamide molecules.

Table 4-1 summarizes the thermodynamic parameters for the dimerization reactions of N-alkylacetamides (Tanford, 1970). Ability to form hydrogen bonds between C=O and N-H groups is lower in chloroform, which acts as a hydrogen-bond donor, and in dioxane, which acts as a hydrogen-bond acceptor, than that in nonpolar solvents such as carbon tetrachloride and benzene. In water $\Delta G_u > 0$ and $\Delta H = 0$. This indicates that the hydrogen bond between a peptide C=O or N-H group and water is stronger than the hydrogen bond between C=O and N-H groups. These findings suggest that the ability to form hydrogen bonds depends on whether the groups involved are located on the surface of the protein molecule and exposed to solvent water, or buried in the hydrophobic interior of the molecule.

When the interior hydrogen bonds are broken and the groups involved are exposed to solvent on unfolding of the protein molecules, these groups are again hydrogen-bonded to water molecules. Thus the interior hydrogen bonds do not appear to contribute to protein stability.

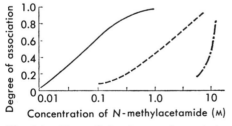

Fig. 4-1. Plots of the degree of association of N-methylacetamide against its concentration in water (—·—), dioxane (·····) and carbon tetrachloride (——). (Klotz, I.M. & Franzen, J.S. (1962) *J. Am. Chem. Soc., 84*, 3461)

TABLE 4-1

Thermodynamic Parameters for Dimerization Reactions of N-alkylacetamides and Related Reactions (25°C)

Solvent	Alkyl group	Association constant (mole fraction unit)	ΔG_u (cal/mol)	ΔH (cal/mol)	ΔS_u (cal/deg/mol)
C_6H_6	Me	69	−2,500	−3,600	−4
	isoPr	10.5	−1,400	—	—
CCl_4	Me	48	−2,350	−5,100	−9
	isoPr	16	−1,650	—	—
$CHCl_3$	Me	13	−1,500	—	—
	isoPr	4.5	−900	—	—
	Bu	3	−650	—	—
Dioxane	Me	6.1	−1,070	−800	+1
	isoPr	3.5	−750	—	—
Dimethyl sulfoxide	isoPr	0.65	+250	—	—
H_2O	Me	0.28	+750	0	−2
⟨Related reactions in water⟩					
Urea + urea		2.2	−400	−1,400	−3
Urea + one peptide		0.7	+200	—	—
Urea + two peptide groups		7.5	−1,200	—	—

Tanford, C. (1970) *Adv. Protein Chem.*, 25, 1.

The free energy changes for hydrogen bond formation between C=O and N–H groups in water and carbon tetrachloride in the following cycle have been estimated (Klotz & Farnham, 1968; Kresheck & Klotz, 1969; Roseman, 1988).

$$(N-H, O=C)_{H_2O} \xrightarrow{(3)} (N-H\cdots O=C)_{H_2O}$$
$$(III) \qquad\qquad (IV)$$
$$(2)\uparrow \qquad\qquad \downarrow (4)$$
$$(N-H, O=C)_{CCl_4} \xleftarrow{(1)} (N-H\cdots O=C)_{CCl_4}$$
$$(II) \qquad\qquad (I)$$

In this cycle, the free energy changes for reactions (1), (2), and (3) were found to be $\Delta G°_1 = +2.4$ kcal/mol, $\Delta G°_2 = −6.12$ kcal/mol and $\Delta G°_3 = 3.1$ kcal/mol. Using the relation $\Delta G°_4 = −(\Delta G°_1 + \Delta G°_2 + \Delta G°_3)$, $\Delta G°_4$ was calculated to be −0.62 kcal/mol. The free energy change for transfer of a group with the hydrogen bond C=O$\cdots$H–O from water to benzene was also small, +0.55 kcal/mol. Thus the stability of these hydrogen-bonded groups is not affected appreciably by solvent polarity. This also suggests that hydrogen bonds do not contribute to protein stability.

If the hydrogen bonds do not contribute significantly to protein stability, what is their role? The linear array of partial charges of three atoms such as N–H···O is important for the formation of a stable hydrogen bond. This geometrical restriction may confer specificity upon interactions between different elements of the protein molecule. Alternatively, the hydrogen bonds may play a part in restricting any fluctuations of the protein molecule (Chapter 6).

On the other hand, it has been reported that the hydrogen bonds between peptide groups are important in stabilizing the helices of poly-L-glutamic acid, poly-L-lysine and poly-L-leucine in water. In these helices, bulky side chains might protect the hydrogen bonds. Similarly, the higher the content of guanine and cytosine in DNA, the higher the melting temperature for the thermal transition. This is because three hydrogen bonds are formed between guanine and cytosine, compared with only two between adenine and thymine. This shows the importance of the hydrogen bonds for DNA stability.

Creighton (1983) pointed out that the effective concentration should be taken into account when interpreting protein stability. The above-mentioned association constant determined using model compounds such as N-methylacetamide is the equilibrium constant for the intermolecular bimolecular reaction. In the protein molecule, however, two interacting groups, A and B, are linked by a single chain and thus the association reaction between A and B is monomolecular. Furthermore, many interactions of various functional groups may occur simultaneously.

The intramolecular and intermolecular interactions between A and B are expressed by Eqs. (4.2) and (4.3), respectively.

$$A - B \overset{K_{unl}}{\rightleftharpoons} A \cdot B \tag{4.2}$$

$$A + B \overset{K_{AB}}{\rightleftharpoons} A \cdot B \tag{4.3}$$

The effective concentration (A/B) of A—B is expressed by K_{uni}/K_{AB}. The equilibrium constant (K_{uni}) for the intramolecular unimolecular reaction is expressed by

$$K_{uni} = K_{AB} \ (A/B). \tag{4.4}$$

The association constants for the formation of hydrogen bonds obtained using model compounds lie within the range 0.005–0.04 M^{-1} in water. As will be described in Section 4-3, the equilibrium constants for the formation of ion-pairs are also only 0.3–0.5 M^{-1}. This suggests that the hydrogen bonds and ion-pairs are not stable in water. However, these interactions do, in fact, exist in the protein molecule. In order for these interactions to be stable, the association constants must be of the order of at least 10^2 M^{-1}. When two interacting groups are far apart, the effective concentration is extremely low. However, when two interacting groups are located at appropriate positions on a flexible chain, the effective concentration is high, and thus although interaction between these two groups estimated when they are present separately may be weak, the association constant K_{uni} will be large and the interaction will occur.

Creighton (1983) considered the following reaction in which two pairs of interactions occur simultaneously and any one of the two pairs increases the effective concentration of the other pair.

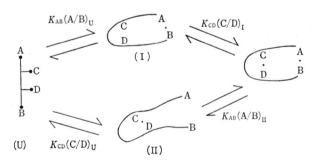

In this scheme, $(A/B)_U$ and $(C/D)_U$ represent the effective concentrations of the pairs A and B, and C and D, respectively, in the unfolded state, and $(C/D)_I$ and $(A/B)_{II}$ represent the effective concentrations of pair C and D in conformation I and pair A and B in conformation II, respectively.

The following relation holds for this cycle

$$\frac{(A/B)_{II}}{(A/B)_U} = \frac{(C/D)_I}{(C/D)_U} = K_{COOP}. \tag{4.5}$$

This indicates that the interaction between C and D in conformation I or the interaction between A and B in conformation II becomes more stable by a factor of K_{COOP} compared with the interaction between C and D or A and B in the unfolded state.

This relation may be extended to the situation within a protein in which many groups can interact simultaneously. The equilibrium constant K_{net} for the reaction from the unfolded molecule in which no interactions exist to the folded molecule in which all groups involved interact completely may be written as follows

$$K_{net} = [K_{AB}(A/B)_U][K_{CD}(C/D)_I][K_{EF}(E/F)_{III}]\cdots, \tag{4.6}$$

where K_{net} is independent of the folding pathway.

This relation can illustrate how sufficient interactions for stabilizing the protein molecule can be obtained from only weak interactions. Figure 4-2 is constructed assuming that the interaction of any one pair among ten weak interactions increases the interaction of the next pair by a factor of 10. In this figure, K_i represents the equilibrium constant for a pair such as K_{AB} or K_{CD}, and $C_{eff,i}$ represents the effective concentration of the ith pair. If the interaction $(K_1 C_{eff,1})$ of the first pair in the unfolded state is assumed to be 10^{-4} M^{-1}, then $K_1 C_{eff,i}$ increases to 10^{-3} by the second interaction, to 10^{-2} by the third interaction and to 10^{-1} by the fourth interaction. Correspondingly, the K_{net} value decreases to 10^{-7}, to 10^{-9} and to 10^{-10} through the second, third, fourth and fifth interactions. From the sixth pair interactions, the K_{net} value begins to increase and becomes 10^5 when all ten interactions are complete.

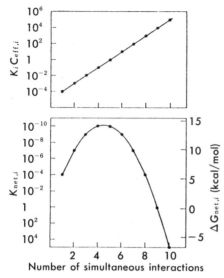

Fig. 4-2. Illustration of the cooperativity of folding produced by 10 weak interactions. (Creighton, T.E. (1983) *Biopolymers, 22*, 49)

This is a value sufficient for stabilizing the protein molecule. Thus protein stability can be explained in terms of weak interactions if the effective concentrations are taken into account.

4-2. HYDROPHOBIC INTERACTIONS

Hydrophobic side chains avoid contact with water and tend to assemble with each other. This type of interaction between hydrophobic residues in water is termed hydrophobic bonding or hydrophobic interaction. The term "hydrophobic bond" was first proposed by W. Kauzmann in 1954. He further analysed the thermodynamic parameters of solution of hydrocarbons in water and non-polar solvents and concluded that the most stable conformation of the protein molecule is one in which the hydrophobic residues are buried in the interior of the molecule and out of contact with water (Kauzmann, 1959). This provided the thermodynamic basis for the "oil drop" model of the protein molecule.*

The free energy change of transfer of a liquid hydrocarbon to water is given by the following equation (see Eq. (1.18))

$$\Delta G_t = -RT \ln X, \tag{4.7}$$

where X is the solubility of the hydrocarbon in water expressed in terms of the mole fraction.

The entropy change of transfer is given by

$$\Delta S° = \Delta H°/T + R \ln X, \tag{4.8}$$

where $\Delta H°$ is the standard enthalpy change of transfer.

Table 4-2 shows the thermodynamic parameters of transfer of liquid hydrocarbons to water at 25°C. The free energy changes of transfer are positive, indicating that the hydrocarbons have no tendency to transfer to water. The positive free energy changes result from the negative entropy changes, which exceed the negative enthalpy changes. Since the parameters given in this table are expressed in terms of unitary units (see Section 1-2), this negative entropy change reflects only the change in the interactions between the solute and solvent and does not include the entropy of mixing. This negative entropy change has been assumed to be due to a change from a pure water structure to a more regular structure of water around the dissolved hydrocarbon molecule. Another characteristic observed for dissolution of hydrocarbons in water is an abnormally large increase in heat capacity. The structure of water surrounding the hydrocarbon molecule has a larger heat capacity than pure liquid water. When the protein molecule is unfolded, an

---

*Klotz (1958) was the first to emphasize the importance of the water structure in the stabilization of protein structure. He estimated the free energy of interaction between non-polar groups and water to be negative, assuming that the interaction enthalpy is approximated by the enthalpy of formation of clathrate hydrates of Ar or Kr (-15 kcal/ mol) and that the entropy change is negative. This led him to conclude that hydrophobic side chains are located on the surface of the protein molecule, and that an ice-like lattice structure of water is formed around the exposed hydrophobic side chains. He inferred that this water structure is important for stabilizing the protein molecule. However, this conclusion does not match the present picture of the protein molecule.

increase in heat capacity is observed, originating from contact of hydrophobic residues exposed to solvent water on unfolding.

The conclusion drawn from the thermodynamic parameters for dissolution of hydrocarbons in water at around room temperature is that the driving force for transfer of hydrophobic side chains from water to the hydrophobic interior of the protein molecule is entropic. In this process the regular structure of water around the hydrophobic residues returns to the structure of pure liquid water.

The decrease in entropy observed when hydrocarbons are dissolved in water is explained as follows. When a solute molecule is dissolved in water, a cavity into which the solute molecule will fit must be formed by separating the strong mutual interactions of the water molecules. For the formation of such a cavity, a large free energy will be required. When a solute molecule which contains a group capable of forming a hydrogen bond, such as an OH group, is placed in the cavity, the unfavourable free energy of the system will be compensated by the formation of solute-water hydrogen bonds. However, when a non-polar solute molecule is placed in the cavity, compensatory hydrogen bonds with water are not formed. Consequently, compensation of the unfavourable free energy will be achieved by reorientation of the water molecules themselves surrounding the solute and by formation of the maximum possible number of hydrogen bonds of water. This is the molecular interpretation for the unfavourable entropy of solution of non-polar solutes in water (Jencks, 1969).

TABLE 4-2
Thermodynamic Parameters of Transfer of Liquid Hydrocarbons to Water (25°C)

Hydrocarbon	Solubility in water (mole fraction $\times 10^{-4}$)	ΔG_u (cal/mol)	ΔH (cal/mol)	ΔS_u (cal/deg/mol)	ΔC_p (cal/deg/mol)
Benzene	4.01	4,590	500	−14	54
Toluene	1.01	5,390	410	−17	63
Ethyl benzene	0.258	6,190	480	−19	76
Cyclohexane	0.117	6,660	−24	−23	86
Pentane	0.095	6,780	−480	−25	96
Hexane	0.020	7,690	0	−26	105

Baldwin, R.L. (1986) *Proc. Natl. Acad. Sci. U.S.A., 83*, 8069.

X-ray crystallographic studies on the water structure surrounding nonpolar side chains have been reported only for crambin (Teeter, 1984, 1989), although there have been several detailed reports on the water structure surrounding the polar or ionizable side chains of proteins. In the α-helix region of crambin, a pentagonal water cluster surrounded by Leu 18, Asn 14, Ile 7 and Val 8, which is very similar to the clathrate hydrate pentagonal ring of small molecules, was observed.

The heat capacity changes for dissolution of hydrocarbons in water are constant within the temperature range 15°C to 35°C. Under these conditions, the temperature dependence of $\Delta H°$ and $\Delta S°$ may be expressed by the following equations

$$\Delta H°(T_2) = \Delta H°(T_1) + \Delta C_p(T_2 - T_1) \tag{4.9}$$

$$\Delta S°(T_2) = \Delta S°(T_1) + \Delta C_p \ln(T_2/T_1). \tag{4.10}$$

Since the entropy change is negative and the heat capacity change is positive for dissolution of hydrocarbons in water (Table 4-2), the absolute value of the entropy change should decrease and approach zero with increasing temperature.

The following empirical rule has been obtained for the transfer of liquid hydrocarbons to water at 25°C (Sturtevant, 1977)

$$\Delta S_u/\Delta C_p = -0.263 \pm 0.046. \tag{4.11}$$

If the temperature at $\Delta S_u = 0$ is denoted by T_s, the following equation is derived from Eq. (4.10)

$$-\Delta S_u/\Delta C_p = \ln(T_s/T). \tag{4.12}$$

If T_s is constant, $\Delta S_u/\Delta C_p$ will be constant. Surprisingly, the T_s values obtained using the values of ΔS_u and ΔC_p for the six hydrocarbons shown in Table 4-2 are found to be the same (386.0 K (112.8° ± 2.4°C)). The same T_s value is also obtained for solution of benzene in water (Table 4-3) (Baldwin, 1986).

If the temperature at $\Delta H° = 0$ is denoted by T_h, the following equation is obtained

$$\Delta H°(T) = \Delta C_p(T - T_h). \tag{4.13}$$

TABLE 4-3

Thermodynamic Parameters for Solution of Benzene in Water at Various Temperatures

Temperature (°C)	Solubility (mole fraction $\times 10^{-4}$)	ΔH (cal/mol)	ΔS_u (cal/deg/mol)	ΔG_u (cal/mol)	T_s (°C)
15	3.99	−36	−16	4,480	112.5
25	4.01	497	14	4,630	112.8
30	4.09	755	13	4,700	113.3
35	4.20	1,044	12	4,790	112.5
				⟨mean⟩	112.8±0.4

Baldwin, R.L. (1986) *Proc. Natl. Acad. Sci. U.S.A.*, **83**, 8069.

The T_h value obtained using this relation is not as constant as that observed for T_s, being 22.2°±5.5°C.

Table 4-3 gives the thermodynamic parameters for dissolution of benzene in water within the temperature range 15° to 35°C. With increasing temperature, ΔS_u approaches zero and ΔG_u increases. This table also shows that the transfer of benzene to water is an entropy-driven process at 15.8°C, but changes to an enthalpy-driven process with increasing temperature.

Very recently, Privalov and Gill (1988) published a review on the thermodynamic behaviour of hydrocarbon dissolution in water. They claimed that hydrophobic interactions lead to destabilization of globular proteins, and that the hydration effect of nonpolar solutes stabilizes the dissolved state (Privalov & Gill, 1989; Privalov, 1989; Murphy et al., 1990). However, this new view of the hydrophobic effect is not generally accepted, and there are arguments against it (Dill, 1990).

The relation between the hydrophobicity (Δg_t) (Section 1-2) and the accessible surface area of amino acid side chains is shown in Fig. 4-3. A good straight line is obtained for Ala, Val, Leu and Phe, for which the slope is 22 cal/mol/Å^2. A good straight line is also obtained for Ser, Thr and Tyr, which have an OH group in the side chain, and its slope is found to be 26 cal/mol/Å^2. When compared for the same accessible surface area, the Δg_t value for the side chain with an OH group is lower by approximately 1 kcal than that for a nonpolar side chain. This is due to possible hydrogen bonding between the OH group and water. This shows that if a

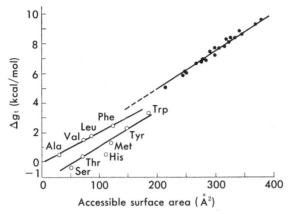

Fig. 4-3.   The relation between hydrophobicity (Δg_t) and accessible surface area.
Small black circles represent the data obtained for hydrocarbons. The straight line
is drawn assuming that the slope is 25 cal/mol/Å². (Richards, F.M. (1977) *Annu.
Rev. Biophys. Bioeng.*, *6*, 151)

correction of 1–2 kcal/mol is made for the Δg_t value, the hydrophobicity of the polar side chain changes to approximately that of
the non-polar side chain. A pair of hydrogen bonding groups can be
treated as a hydrophobic unit, buried in the hydrophobic interior
of a protein molecule (see Section 4-1). The amino acid residue has
a hydrophobicity of 24 cal/mol/Å² on average. Figure 4-3 also
shows the relation between the free energy change of transfer of
hydrocarbons to water and the accessible surface area. The slope is
found to be 20–28 cal/mol/Å² depending on the estimate of the
accessible surface area. On average, a value of 24 cal/mol/Å² may
be used to estimate the contribution of the hydrophobic free energy
change.

This value of hydrophobicity can be used to estimate the free
energy change due to hydrophobic interactions of folding from the
unfolded to native state. Using Eqs. (3.1) and (3.2) for monomeric
proteins or Eqs. (3.4) and (3.2) for oligomeric proteins, the decrease
in the accessible surface area on folding can be evaluated. For
instance, this decrease in the accessible surface area is 14,080 Å² for
hen egg-white lysozyme and 5,740 Å² for bovine pancreatic trypsin
inhibitor. Thus the hydrophobic free energy change of folding is

calculated to be $24\,\mathrm{cal/mol} \times 14{,}080 = 338\,\mathrm{kcal/mol}$ for lysozyme and $24\,\mathrm{cal/mol} \times 5{,}740 = 138\,\mathrm{kcal/mol}$ for trypsin inhibitor. The contribution of the hydrophobic free energy is thus considerable. Since lysozyme and trypsin inhibitor consist of 129 and 58 amino acid residues, respectively, the hydrophobic free energy per residue amounts to approximately $2.5\,\mathrm{kcal/mol}$.

4-3. ELECTROSTATIC INTERACTIONS

The ionization behaviour of ionizable side chains has been described in Section 1-1. Many ionizable residues exist in the protein molecule, and the electrostatic interactions of these groups are naturally important for stabilizing the protein molecule. However, estimation of the contribution of electrostatic interactions to protein stability is still equivocal. When two charges q_i and q_j are separated by a distance r, the electrostatic free energy may be expressed as $q_i q_j / Dr$, where D, the dielectric constant, depends on the medium in which the charges are immersed. When two charges are separated by 3 Å, the electrostatic free energy between them is calculated to be $110\,\mathrm{kcal/mol}$ in a vacuum ($D=1$) and $110/80$ kcal/mol in water ($D=80$). In the protein molecule, the dielectric constant of the medium surrounding an ionizable group is not known, and has been estimated to range from 5 to 50 depending on the environment. Thus it is difficult to calculate the electrostatic free energy change for ionizable groups in proteins.

In the following treatment, we first consider the ionization equilibria of simple monobasic acids, and then the ionization behaviour and electrostatic interactions of ionizable groups on the protein molecule.

Consider the ionization equilibrium of a monobasic acid AH,

$$\mathrm{AH} = \mathrm{A}^- + \mathrm{H}^+ \tag{4.14}$$

for which the ionization constant K is

$$K = \frac{(\mathrm{A}^-)(\mathrm{H}^+)}{(\mathrm{AH})}. \tag{4.15}$$

The fraction ($\bar{r}$) of the concentration of the ionized form ((A^-)) to the total concentration of the acid (($A)_0$) is expressed by

$$\bar{r} = \frac{(A^-)}{(A)_0} = \frac{K/(H^+)}{1 + K/(H^+)}. \tag{4.16}$$

The ionization equilibria of a dibasic acid AH_2 are

$$AH_2 = AH^- + H^+, \quad K_1 = \frac{(AH^-)\,(H^+)}{(AH_2)} \tag{4.17}$$

$$AH^- = A^{--} + H^+, \quad K_2 = \frac{(A^{--})\,(H^+)}{(AH^-)}. \tag{4.18}$$

K_1 and K_2 are the ionization constants of AH_2 and AH^-, respectively.

If the two ionizable groups of a dibasic acid are the same, and the two protons dissociate independently, then the ionization equilibria of the dibasic acid may be expressed by

$$
\begin{array}{ccc}
 & \text{H-A-H} & \\
K_{\text{int}} \nearrow & & \nwarrow K_{\text{int}} \\
{}^-\text{A-H} & & \text{H-A}^- \\
K_{\text{int}} \searrow & & \swarrow K_{\text{int}} \\
 & \text{A}^{--} &
\end{array}
\tag{4.19}
$$

where K_{int} is the intrinsic ionization constant.

For the ionization equilibrium

$$\text{H-A}^- = \text{A}^{--} + H^+, \tag{4.20}$$

the ionization constant K_{int} is

$$K_{\text{int}} = \frac{(A^{--})\,(H^+)}{(\text{H-A}^-)}. \tag{4.21}$$

For the ionization equilibrium

$$^-\text{A-H} = \text{A}^{--} + H^+, \tag{4.22}$$

the ionization constant K_{int} is

$$K_{\text{int}} = \frac{(A^{--})\,(H^+)}{(^-\text{A-H})}. \tag{4.23}$$

The concentration of the ionized form AH^- is the sum of the concentrations of ^-A-H and $H-A^-$

$$(AH^-)=(^-A-H)+(H-A^-). \qquad (4.24)$$

Therefore, the equilibrium constant K_2 for the ionization of AH^- is given by

$$K_2=\frac{(A^{--})\,(H^+)}{(^-A-H)+(H-A^-)}=\frac{K_{\text{Int}}}{2}. \qquad (4.25)$$

Similarly, for the ionization equilibrium

$$H-A-H=H-A^-+H^+, \qquad (4.26)$$

the ionization constant K_{Int} is

$$K_{\text{Int}}=\frac{(H-A^-)\,(H^+)}{(H-A-H)}. \qquad (4.27)$$

For the ionization equilibrium

$$H-A-H=^-A-H+H^+, \qquad (4.28)$$

the ionization constant K_{Int} is

$$K_{\text{Int}}=\frac{(^-A-H)\,(H^+)}{(H-A-H)}. \qquad (4.29)$$

Therefore, the equilibrium constant K_1 for the ionization of $H-A-H$ is given by

$$K_1=\frac{\{(H-A^-)+(^-A-H)\}(H^+)}{(H-A-H)}=2K_{\text{Int}}. \qquad (4.30)$$

Thus the ratio of the two macroscopic constants K_1/K_2 equals 4. This factor is called the statistical factor for the ionization of dibasic acids. Even if the two groups are the same and the protons dissociate independently, the macroscopic ionization constant K_1 for the first ionization is greater by a factor of 4 than the macroscopic ionization constant K_2 for the second ionization.

For the ionization of succinic acid, $\begin{smallmatrix} CH_2COOH \\ | \\ CH_2COOH \end{smallmatrix}$ $K_1=8.71\times$

10^{-5} and $K_2 = 4.77 \times 10^{-6}$; hence $K_1/K_2 = 18.3$. This is far from the statistical factor 4 for the ionization of the dibasic acid. This indicates that there are electrostatic interactions between the two carboxyl groups. When one of the two carboxyl groups is first ionized, the other carboxyl is unionized. When a proton has been lost from one carboxyl group, however, it becomes more difficult to remove a proton from the second carboxyl group, because the negative charge of AH^- tends to hold the proton more tightly. When the two carboxyl groups are not far apart in the dicarboxylic acids $HOOC(CH_2)_nCOOH$, they will not ionize independently due to electrostatic interaction. As the distance between the two groups increases, the electrostatic interaction between them will become smaller and the value of K_1/K_2 will approach the statistical factor 4. In fact, when $n=1$, $K_1/K_2 = 650$, and even when $n=7$, the value of K_1/K_2 is still 6.5 (Fig. 4-4).

Consequently, the ionization of dibasic acids must be expressed by the following scheme in which the microscopic ionization constants (k_1, k_2, k_3 and k_4) of the respective ionization processes are not the same

$$
\begin{array}{ccc}
 & H\text{-}A\text{-}H & \\
k_1 \nearrow & & \nwarrow k_2 \\
{}^-AH & & HA^- \\
k_3 \searrow & & \swarrow k_4 \\
 & A^{--} &
\end{array}
\tag{4.31}
$$

Since the microscopic ionization constants are expressed by the following equations

$$k_1 = \frac{(^-AH)\,(H^+)}{(HAH)}, \quad k_2 = \frac{(HA^-)\,(H^+)}{(HAH)}$$

$$k_3 = \frac{(A^{--})\,(H^+)}{(^-AH)}, \quad k_4 = \frac{(A^{--})\,(H^+)}{(HA^-)}. \tag{4.32}$$

the following relation is obtained

$$k_1 k_3 = k_2 k_4. \tag{4.33}$$

When a dibasic acid is titrated with an alkali metal, the average number of protons released ($\bar{h}$) is given by

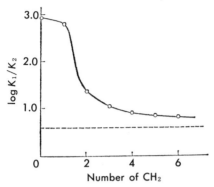

Fig. 4-4. The values of K_1/K_2 as a function of the number of CH_2 (n) of dibasic acids $HOOC(CH_2)_nCOOH$. (Bull, H.B. (1964) "An Introduction to Physical Biochemistry," F.A. Davis Co., Philadelphia, p. 117)

$$\bar{h}=\frac{(^-AH)+(HA^-)+2(A^{--})}{(HAH)+(^-AH)+(HA^-)+(A^{--})}=\frac{\dfrac{k_1+k_2}{(H^+)}+\dfrac{2k_1k_2}{(H^+)^2}}{1+\dfrac{k_1+k_2}{(H^+)}+\dfrac{k_1k_2}{(H^+)^2}} \quad (4.34)$$

For HAH, $h=0$, for ^-AH and HA^-, $h=1$, and for A^{--}, $h=2$. The value of h changes from 0 to 2 as the titration proceeds from strongly acid to strongly alkaline solutions.

If we consider a mixture of two simple monobasic acids, HA_1 and HA_2, for which the ionization constants are G_1 and G_2, respectively, then

$$G_1=\frac{(H^+)\,(A_1^-)}{(HA_1)}, \quad G_2=\frac{(H^+)\,(A_2^-)}{(HA_2)}. \quad (4.35)$$

When the mixture is titrated with an alkali metal, the average number of protons released ($\bar{h}$) is expressed by the equation

$$\bar{h}=\frac{\dfrac{G_1}{(H^+)}}{1+\dfrac{G_1}{(H^+)}}+\frac{\dfrac{G_2}{(H^+)}}{1+\dfrac{G_2}{(H^+)}}=\frac{\dfrac{G_1+G_2}{(H^+)}+\dfrac{2G_1G_2}{(H^+)^2}}{1+\dfrac{G_1+G_2}{(H^+)}+\dfrac{G_1G_2}{(H^+)^2}}. \quad (4.36)$$

Equation (4.36) is identical in form to Eq. (4.34). Comparison of the two equations gives the following relations

$$\left.\begin{array}{l}G_1+G_2=k_1+k_2 \\ G_1G_2=k_1k_3=k_2k_4.\end{array}\right\} \quad (4.37)$$

G_1 and G_2 are referred to as the titration constants (Edsall & Wyman, 1958). Here, we have examined the titration constants of dibasic acids. In general, the titration curve of a polybasic acid containing n ionizable groups can be described as if it were the titration curve of a mixture of n simple monobasic acids with suitably chosen ionization constants (titration constants).

When the titration curve of a dibasic acid is given, we can choose G_1 and G_2, which fit Eq. (4.36). If any one of the four microscopic ionization constants is known, the other three can be determined.

As described above, the macroscopic ionization constants, K_1 and K_2, of dibasic acids are given by

$$K_1 = \frac{(\mathrm{H^+})\{(\mathrm{HA^-}) + (^-\mathrm{AH})\}}{(\mathrm{HAH})} = k_1 + k_2 \qquad (4.38)$$

$$\left. \begin{array}{l} K_2 = \dfrac{(\mathrm{H^+})\,(\mathrm{A^{--}})}{\{(\mathrm{HA^-}) + (^-\mathrm{AH})\}} \\[2mm] \dfrac{1}{K_2} = \dfrac{1}{k_3} + \dfrac{1}{k_4}. \end{array} \right\} \qquad (4.39)$$

Then

$$K_1 K_2 = k_1 k_3 = k_2 k_4. \qquad (4.40)$$

Therefore, the macroscopic ionization constants are related to the titration constants by

$$\left. \begin{array}{l} K_1 = G_1 + G_2 \\ K_1 K_2 = G_1 G_2. \end{array} \right\} \qquad (4.41)$$

Thus the values of K_1 and K_2 can be evaluated if the values of G_1 and G_2 are known.

The titration constants are useful not only for analysing the titration curves of polybasic acids, but also for determining the pK_a values of two catalytic groups located close together on an enzyme. The latter will be examined using hen egg-white lysozyme as an example in Chapter 8.

Generally, when n identical ionizable groups of a polymer ionize independently, the fraction $(\bar{F})$ of the ionized form is given by

$$\bar{F} = \frac{nK_{\text{int}}/(H^+)}{1 + K_{\text{int}}/(H^+)}. \qquad (4.42)$$

Any of the n macroscopic constants (for instance, K_j) is expressed by

$$K_j = \left(\frac{n-j+1}{j}\right) K_{\text{int}}, \qquad (4.43)$$

where $(n-j+1)/j$ is the statistical factor.

Ionizable groups of proteins include carboxyl, amino, imidazole, phenolic hydroxyl, sulfhydryl and guanidyl groups, with different pK_a values. When the number of ionizable groups of each type is denoted by n_1, n_2, $\cdots\cdots$ and the various ionizable groups ionize independently with the intrinsic ionization constants $K_{1,\text{int}}$, $K_{2,\text{int}}$, $\cdots\cdots$, the fraction of the ionized form is expressed by

$$\bar{F} = \frac{n_1 K_{1,\text{int}}/(H^+)}{1 + K_{1,\text{int}}/(H^+)} + \frac{n_2 K_{2,\text{int}}/(H^+)}{1 + K_{2,\text{int}}/(H^+)} + \cdots\cdots. \qquad (4.44)$$

The following equations are obtained for the ionization of ionizable group i

$$\frac{\bar{F}_i}{n_i - \bar{F}_i} = \frac{x_i}{1 - x_i} = \frac{K_{i,\text{int}}}{(H^+)} \qquad (4.45)$$

$$pH - \log\frac{x_i}{1 - x_i} = pK_{i,\text{int}}, \qquad (4.46)$$

where x_i is the degree of ionization. Equation (4.46) is the same as the Henderson-Hasselbalch equation, which describes the ionization of simple monobasic acids.

As will be described in Chapter 5, almost all the proteins assume a randomly coiled conformation in 6 M guanidine hydrochloride. In a medium of high ionic strength such as 6 M guanidine hydrochloride, the electrostatic effect of the molecule as a whole and the interactions of ionizable groups of the randomly coiled protein molecule are assumed to be extremely small, so that each ionizable group will be titrated independently. Thus the titration curve of a protein in 6 M guanidine hydrochloride may be analysed

TABLE 4-4

Analysis of Titration Curves of Ribonuclease A and Hen Egg-white Lysozyme in 6 M Guanidine Hydrochloride

Ionizable group		Ribonuclease A		Lysozyme	
		Number of residues used for calculation	pK_{int}	Number of residues used for calculation	pK_{int}
	α-COOH	1	3.4	1	3.4
Asp	β-COOH	5	3.8	7	3.9
Glu	γ-COOH	5	4.3	2	4.35
His		4	6.5	1	6.5
Tyr		{ 3	9.75	3	9.9
		{ 3	10.5		
	α-NH$_2$	1	7.6	1	7.6
Lys	ε-NH$_2$	10	10.35	6	10.35
Arg		4	>12	1	>12

Nozaki, Y. & Tanford, C. (1967) *J. Am. Chem. Soc., 89*, 742; Roxby, R. & Tanford, C. (1971) *Biochemistry, 10*, 3348.

using Eq. (4.44). In fact, the titration curves of ribonuclease A and hen egg-white lysozyme in 6 M guanidine hydrochloride have been analysed using Eq. (4.44). The pK_a value of the ionizable group of each type of ribonuclease A agrees well with that of lysozyme (Table 4-4). These pK_a values also agree well with the values obtained for small model compounds (Table 1-2).

However, ionization of ionizable groups in a native protein molecule requires a more refined treatment, because they lie close together. The simplest analysis can be done by treating the protein molecule as a multivalent impenetrable spherical particle with a net charge ($\bar{Z}$) distributed uniformly on the surface, and assuming that the electrostatic effect due to the net charge affects the release of protons from ionizable groups (Linderstrøm-Lang, 1924). The free energy change of ionization ($\Delta G°$) of an ionizable group will be written as the sum of $\Delta G°_{int}$, which corresponds to the non-cooperative intrinsic ionization constant K_{int}, and the electrostatic free energy of interaction $\Delta G°(\bar{Z})$.

$$\Delta G° = \Delta G°_{int} + \Delta G°(\bar{Z}) \tag{4.47}$$

$$\Delta G°_{int} = -RT \ln K_{int}, \tag{4.48}$$

where $\Delta G^\circ(\bar{Z})$ is the work required to bring in a proton against the electric field produced by the net charge $(\bar{Z})$ on the protein molecule. $\Delta G^\circ(\bar{Z})$ may be approximated as being proportional to $\bar{Z}$:

$$\Delta G^\circ(\bar{Z}) = -2\omega RT\bar{Z}. \tag{4.49}$$

For a spherical protein molecule, ω is estimated using the equation

$$\omega = \frac{\varepsilon^2}{2DkT}\left(\frac{1}{b} - \frac{\kappa}{1+\kappa a}\right), \tag{4.50}$$

where ε is the elementary charge, k is the Boltzmann constant, D is the dielectric constant, κ is the Debye-Hückel parameter (see Eq. (1.11)) and b is the radius of the spherical protein molecule. a is the mean distance of approach of the surrounding ions, and is usually set equal to $b+2.5$ Å.

On the basis of these assumptions, the following equation is obtained

$$\bar{r} = \frac{n_1 K_{1,\text{int}} e^{2\omega\bar{Z}}/(\text{H}^+)}{1+K_{1,\text{int}} e^{2\omega\bar{Z}}} + \frac{n_2 K_{2,\text{int}} e^{2\omega\bar{Z}}/(\text{H}^+)}{1+K_{2,\text{int}} e^{2\omega\bar{Z}}} + \cdots. \tag{4.51}$$

For the ionization of ionizable group i

$$\frac{\bar{r}_i}{n-\bar{r}_i} = \frac{x_i}{1-x_i} = \frac{K_{i,\text{int}} e^{2\omega\bar{Z}}}{(\text{H}^+)} \tag{4.52}$$

$$\text{pH} - \log\frac{x_i}{1-x_i} = pK_{i,\text{int}} - 0.868\,\omega\,\bar{Z}. \tag{4.53}$$

Equation (4.53) is referred to as the Linderstrøm-Lang equation and has been used to analyse the titration curves of proteins.

The Linderstrøm-Lang equation was derived by assuming that charges are distributed uniformly on the spherical protein molecule and that its net charge affects the release of protons. For native proteins, however, the distribution of charges is not uniform. A theoretical treatment of the electrostatic effect which takes this fact into account was developed by Tanford and Kirkwood (1957). They calculated the electrostatic free energy for a set of discrete point-charges on a spherical surface of radius b and ion-exclusion radius a. The positions of charges can be represented by vector r_i

from the centre of the sphere. The charges were treated at the surface of the equivalent sphere, which is assumed to form a continuous medium of low dielectric constant D_i surrounded by solvent water of dielectric constant D. When the electrostatic interaction between two charges at positions i and j is denoted by W_{ij}, the total electrostatic free energy W_{el} is expressed by

$$W_{el} = \sum_{i \neq j} W_{ij} \tag{4.54}$$

$$W_{ij} = \varepsilon^2 Z_i Z_j \left\{ \frac{A_{ij} - B_{ij}}{b} - \frac{C_{ij}}{a} \right\}, \tag{4.55}$$

where ε is the elementary charge, A_{ij}, B_{ij} and C_{ij} are complex functions of a, D, D_i, r_{ij} (the distance between i and j) and the distance between the charge position and the surface of the sphere (d). $(A_{ij} - B_{ij})$ corresponds to the interaction at zero ionic strength and C_{ij} is the term which corrects the interaction at a given ionic strength.

For a fixed charge distribution, the pK_a value of ionizable group i is given by

$$pK_i = pK_{i,\text{int}} - \sum_{j \neq i} \frac{W_{ij}}{2.303 \, Z_i kT}, \tag{4.56}$$

where $pK_{i,\text{int}}$ is the value shown in Table 1-2.

When the values of a, b, D, D_i and d are given, A_{ij}, B_{ij} and C_{ij} are the functions of r_{ij} only. Tanford and Roxby (1972) analysed the titration curve of hen egg-white lysozyme. Using r_{ij} values determined from X-ray crystallographic data, and assuming $b = 17.5$ Å, $a = 20.0$ Å, $D = 78.5$, $D_i = 4.0$ and $d = 0.4$ Å, they obtained the pK values of the acidic ionizable groups. In this treatment, the positions of the charges are fixed at $d = 0.4$ Å. However, the charge positions are, in fact, different, and hence the dielectric constants around the charges also differ. This may affect the estimated values of W_{ij}.

As described above, the dielectric constant of the environment of a charged group is difficult to estimate. Attempts have been made to estimate the electrostatic interaction using the accessible surface area of charged groups instead of the dielectric constant, using the equation (Matthew, 1985)

$$W_{ij}' = W_{ij}(1 - \overline{SA_{ij}}),\qquad(4.57)$$

where $\overline{SA_{ij}}$ is the average of the relative accessibilities of two interacting ionizable groups i and j. When groups i and j are exposed to solvent, and $\overline{SA_{ij}}$ is almost equal to one, they are located in an environment with a large dielectric constant and the electrostatic interaction between them is small. On the other hand, when $\overline{SA_{ij}}$ is small, the groups i *and* j are located in an environment of low dielectric constant, and thus the electrostatic interaction is large.

Figure 4-5 shows the sum of W_{ij}' for all the ion-pairs of myoglobin as a function of pH. The electrostatic free energy of stabilization amounts to approximately 10 kcal/mol at neutral pH.

Not only the electrostatic factors, but various factors such as hydrogen bonds, salt bridges and nearby hydrophobicity also affect the ionization behaviour of ionizable groups. The apparent pK_a

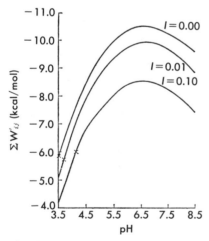

Fig. 4-5. Sum of the electrostatic free energy ($\Sigma W_{ij}'$) for all the ion-pairs of myoglobin as a function of pH. I, ionic strength; x, midpoint of acid denaturation. (Friend, S.H. & Gurd, R.R. (1979) *Biochemistry, 18,* 4612)

TABLE 4-5
The Apparent pK_a Values of the Ionizable Groups of Hen Egg-white Lysozyme (25°C, ionic strength 0.1)

Ionizable group	pK_a	Methods
α-COOH	3.1	NMR
Glu 7	2.6	NMR
Glu 35	6.1	CD, UV, fluorescence, NMR, binding of substrate analogues
Asp 18	(2.0)	Titration
Asp 48	4.3	Binding of substrate analogues
Asp 52	3.4	CD, UV, fluorescence, NMR, binding of substrate analogues
Asp 66	(1.5)	CD, binding of substrate analogues
Asp 87	(3.0)	NMR, titration
Asp 101	4.5	Binding of substrate analogues
Asp 119	(3.0)	Titration
His 15	5.8	NMR, tritium exchange
α-NH$_2$	(7.8)	Titration
Lys 1	10.6	
Lys 13	10.3	
Lys 33	10.4	NMR
Lys 96	10.7	
Lys 97	10.1	
Lys 116	10.2	
Tyr 20	(11.0)	UV
Tyr 23	9.8	CD, NMR
Tyr 53	(11.5)	UV

UV, ultraviolet absorption.

values of the ionizable groups of hen egg-white lysozyme estimated using various methods are summarized in Table 4-5.

It is generally believed that a protein is most stable at its isoelectric point, where the net charge is zero. For instance, ribonuclease T1 is most stable at its isoelectric point, pH 4.5. The free energy changes of unfolding at pH 4.5, where the net charge on the molecule is zero, and at pH 10, where the net charge is about -12, were estimated to be 9 and 3 kcal/mol, respectively (Pace *et al.*, 1990). However, there are many instances of proteins being most stable at pH values considerably different from their isoelectric points. For instance, the thermal stability of T4 phage lysozyme, whose isoelectric point lies above pH 10, is greatest at pH 5. In such a case, contributions other than the total net charge are involved in determining the protein stability (Anderson *et al.*, 1990). For instance, burial of an ionizable group in the hydrophobic interior of the molecule, or formation of an ion-pair between a positive charge and a negative charge will contribute to the protein's stability. If we assume that an ion-pair is formed when the distance between two charges is less than 4 Å, 229 charges are found to be located within this distance range in 38 different proteins. Of these, 85 charges have the same sign and are thus repulsive, and about 85% of the ion-pairs are formed on the surface of the protein molecule. Ion-pairs are not found to be well conservative when corresponding proteins of different origin are compared (Barlow & Thornton, 1983).

Deoxyhaemoglobin has eight ion-pairs, whereas oxyhaemoglobin has none. This is one of the reasons why deoxyhaemoglobin is more stable than oxyhaemoglobin. When chymotrypsinogen is activated with trypsin to yield active chymotrypsin, an ion-pair is formed between the α-NH$_3^+$ of Ile and the β-COO$^-$ of Asp 194. This ion-pair is buried in the interior of the protein molecule and a value of -2.9 kcal/mol is estimated to contribute to the free energy of stabilization. Bovine pancreatic trypsin inhibitor has an ion-pair between the α-NH$_3^+$ of Arg 1 and the α-COO$^-$ of Ala 58. This ion-pair was not revealed by crystallographic analysis, because the study was carried out at pH 10, when there is no charge on the α-amino group of Arg 1. By comparing the stability of the native inhibitor with that of an inhibitor with a modified α-amino group at Arg 1, it was found that the ion-pair contributes to the free energy of stabilization by about 1 kcal/mol. T4 phage lysozyme has an ion-pair between Asp 70 and His 31. The pK_a value of Asp 70 was found to be 0.5 in the folded state and 3.5–4.0 in the unfolded state, and the pK_a value of His 31 was found to be 9.1 in the folded state and 6.8 in the unfolded state. This ion-pair was estimated to stabilize the native state by 3–5 kcal/mol. The dependence of T_m (the transition temperature of heat denaturation) on pH is well explained if the above pK_a values are assumed. The presence of ion-pairs seems to be important for stabilization of thermostable proteins to thermal denaturation. The important role of ion-pairs in stabilization of the helices of the S-peptide and the C-peptide of ribonuclease A was described on p. 58.

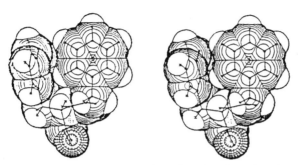

Fig. 4-6. Stereoscopic drawing of N-phenylacetyl-L-phenylalanine. (Burley, S.K. & Petsko, G.A. (1985) *Science, 229*, 23)

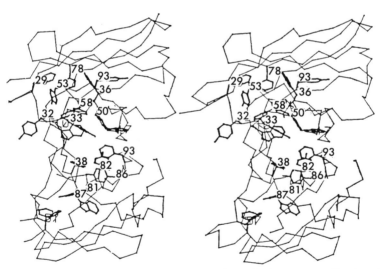

Fig. 4-7. Stereoscopic drawing of the antigen-binding V_L and V_H domains of immunoglobulin Fab fragment. (Burley, S.K. & Petsko, G.A. (1985) *Science, 229*, 23)

4-4. AROMATIC-AROMATIC INTERACTIONS

Recently it has been pointed out that interactions between aromatic side chains contribute to the stability of proteins (Burley & Petsko, 1985, 1988). X-ray crystallographic studies of a peptide containing two phenyl groups and its analogues have shown that the two phenyl groups are arranged face-to-edge (Fig. 4-6). This arrangement is very similar to that found in the crystal structure of benzene. Geometric analysis of the structures of 33 different proteins shows that aromatic groups pair preferentially with centroid separations of between 3.4 and 6.5 Å, and that the angle formed by two aromatic ring planes lies within the range 60° to 90°. The enthalpic contribution of each pair of aromatic-aromatic interactions was estimated to be -1 to -2.5 kcal/mol. As shown in Fig. 4-7, there are many aromatic-aromatic pairs in the antigen-binding domains of the immunoglobulin Fab fragment (Burley & Petsko, 1985). In addition to the aromatic-aromatic interactions, oxygen-aromatic, amino-aromatic and sulphur-aromatic interactions have also been shown to contribute to protein stability.

4-5. CONFORMATIONAL ENTROPY

The various interactions described so far stabilize the native protein molecule. On the other hand, conformational entropy destabilizes the native protein molecule and stabilizes the unfolded protein molecule, because the conformational entropy of the unfolded molecule is much larger than that of the native molecule. The conformational entropy of a randomly coiled polypeptide chain is estimated to be 2–10 e.u. per residue. If the entropy is assumed to be 10 e.u./residue and the freedom of side chains is also taken into account, then the contribution of the entropy term to the free energy amounts to approximately 400 kcal/mol for ribonuclease A and 500 kcal/mol for β-lactoglobulin. Since the folded conformation also has some degree of freedom, the contribution of

TABLE 4-6
Estimated Contributions of Individual Factors to Stability of a Hypothetical Protein of 100 Residues (25°C)

Factor	Free energy difference between N and D (kcal/mol)
Conformational entropy	$-300 - -1,000$
Unfavourable interactions in folded state	-200
Hydrophobic interactions	$+264$
van der Waals interactions	$+227$
Required contribution of hydrogen bonds	$+49 - +719$
Observed net effect	$+10$

N, native state; D, denatured state.
Creighton, T.E. (1983) *Biopolymers, 22*, 49.

the entropy difference between the unfolded and folded states to the free energy will be 200–250 kcal/mol, or about 2 kcal/mol per residue. This destabilizing free energy competes with the sum of the free energies of various factors that stabilize the protein molecule.

In the unfolded state, the conformational entropy of a protein with disulphide bonds is smaller than that of the protein in which the disulphide bonds are reduced. Therefore, the change in the conformational entropy on unfolding of a protein with disulphide bonds is smaller than that of a protein in which the disulphide bonds are reduced, and thus the protein molecule is stabilized by the presence of such bonds. This will be discussed in detail in Chapter 5.

At present, it is difficult to evaluate protein stability quantitatively in terms of the various factors described above. Creighton (1983) has estimated the net of free energy contributions of various factors that stabilize a hypothetical protein consisting of 100 amino acid residues (Table 4-6). In this evaluation, a value of 3.3–10 kcal/mol per residue was used as the contribution of conformational entropy to the free energy. An unfavourable interaction of 2 kcal/mol per residue was assumed for the folded structure. These two factors destabilize the protein. The hydrophobic free energy was calculated assuming 24 cal/mol/Å^2 and using Eqs. (3.1) and (3.2). The van der Waals interaction due to close packing was estimated from the average enthalpy of fusion of small, non-polar model compounds. If the net stabilization free energy is assumed to be + 10 kcal/mol, the hydrophobic and van der Waals interactions are insufficient and other factors must be considered to account for the free energy of +49 to +719 kcal/mol. Analysis of X-ray crystallographic data for various proteins suggests that, on average, 74 hydrogen bonds exist in a protein of 100 residues. If only the hydrogen bonds are considered, a free energy of 0.7-9.7 kcal/mol per residue will be necessary. As described above, however, the intermolecular equilibrium constant for the formation of a hydrogen bond in water is estimated to be only of the order of 10^{-2} M^{-1}. Therefore, we must assume an effective concentration of $310-1.4 \times 10^9$ in order to account for the net free energy of stabilization. Since other factors such as electrostatic and aromatic-aromatic interactions will also be involved in protein stability, evaluation of each factor given in Table 4-6 is only tentative.

5
Stability of Proteins

In Chapter 4 various factors that stabilize the protein molecule were discussed. The regular native protein conformation is stable at physiological pH values, temperatures and salt concentrations, and is easily destroyed at extremely acidic and alkaline pH and high temperatures. Addition of denaturants such as urea, guanidine hydrochloride, organic solvents, salts or surface-active agents also unfolds the protein molecule. The conformation of the protein molecule is also changed at the air-water or oil-water interface. These phenomena are called denaturation (or unfolding) of the protein. Denaturation is defined as changes in protein conformation caused by changes in non-covalent interactions without any accompanying change in covalent interactions.* This definition is said to have been first presented by A.E. Mirsky and L. Pauling (1936). However, five years earlier in 1931, H. Wu gave almost the same definition in Chinese Journal of Physiology: "The compact and crystalline structure of the normal protein molecule, being

---

*Care has to be exercised in studies of alkaline denaturation, since breakage of peptide bonds or disulphide bonds by alkali is frequently observed.

formed by virtue of secondary valences, is easily destroyed by physical as well as chemical forces. Denaturation is disorganization of the natural protein molecule, the change from the regular arrangement of a rigid structure to the irregular diffuse arrangement of the flexible open chain". This definition is particularly remarkable, since at the time the properties of the protein molecule were not well known.

People seem to have been aware of the phenomenon of protein denaturation since the time when proteins were first discovered. Since it was found that protein functions are lost on exposure to extreme pH or high temperature, denaturation has been studied actively from a practical viewpoint. From a purely scientific standpoint, however, protein denaturation has also been an important subject. As early as 1929, on the basis of observations indicating that denaturation might be a reversible process, Anson and Mirsky investigated the thermodynamic parameters of protein denaturation. It was pointed out that an extremely large increase of entropy accompanies denaturation and that the denaturation reaction is an endothermic process. However, because the properties of the protein molecule were not understood well enough and because denaturation behaviour differs greatly between proteins, it was difficult to understand the denaturation reactions in terms of molecular events. When the structures of proteins were first elucidated by X-ray crystallography in 1960, interest and studies on denaturation intensified further. There are two purposes to the study of denaturation. One is to clarify the mechanism of stabilization of the protein molecule by observing the behaviour of proteins under various denaturing conditions. This is the main concern of this chapter, although it was discussed in part in the previous chapter. The other is to clarify the pathway of folding by studying the process of refolding from the unfolded molecule under physiological conditions. Such studies should yield clues to understanding the mechanism by which the native conformation is acquired once the polypeptide chain has been synthesized *in vivo*. This will be described in Chapter 7.

5-1. DENATURED STATE

The native state of the protein molecule can be assumed to be the structure determined by X-ray analysis. However, the denatured state depends upon the denaturing conditions, such as pH, heat and high concentrations of denaturing agents. The protein conformation is greatly changed and assumes a form very similar to a random coil in the presence of urea or guanidine hydrochloride at high concentrations. On the other hand, the content of hydrogen bonds increases relative to that of the native protein upon denaturation by organic solvents such as alcohols. Acid or heat denaturation causes a great conformational change, but water is a poor solvent for denatured proteins and thus the change in molecular dimensions is not as great as that observed for denaturation by urea or guanidine hydrochloride.

The presence of guanidine hydrochloride at high concentrations changes the conformation of most proteins to a random coil. This has been demonstrated by measurements of hydrodynamic properties such as viscosity and sedimentation constant (Tanford *et al.*, 1967; Lapanje & Tanford, 1967), acid-base titration (see Section 4-3) and optical rotation (Tanford *et al.*, 1967) in 6 M guanidine hydrochloride. The intrinsic viscosity $[\eta]$, which is a measure of the effective hydrodynamic volume per gram of the molecule and is related to the compactness of the molecule, is very small, about 3 ml/g, for most proteins under native conditions, but is very large in 6 M guanidine hydrochloride (Table 5-1). Disulphide bonds restrict the dimensions of the protein molecule. On reduction of the disulphide bonds with an SH compound such as β-mercaptoethanol, the intrinsic viscosity of proteins in 6 M guanidine hydrochloride increases further, indicating an additional increase in molecular dimensions. The relationship between the value of log $[\eta]$ and the logarithm of the number of residues (n) for reduced proteins is linear (Fig. 5-1) and is expressed by the equation (Tanford *et al.*, 1967)

$$[\eta] = 0.716\ n^{0.66}. \qquad\qquad (5.1)$$

This is the relation expected for randomly coiled high-polymer chains and indicates that most proteins whose disulphide bonds are reduced behave as random coils in 6 M guanidine hydrochloride. Furthermore, in this solvent, the intrinsic viscosities of proteins with intact disulphide bonds are smaller than those of proteins with reduced disulphide bonds, indicating that the disulphide bonds

TABLE 5-1
Intrinsic Viscosity ($[\eta]$) of Native and Denatured Proteins

Protein	Number of S–S bonds per 100 residues	$[\eta]$ (ml/g)		
		Native	Denatured (6 M guanidine hydrochloride)	Proteins with disulphide bonds broken (6 M guanidine hydrochloride)
Ribonuclease A	3.2	3.3	9.4	16.0
Lysozyme	3.1	2.7	6.5	17.1
Serum albumin	2.8	3.7	22.0	52.2
Chymotrypsinogen	2.0	2.5	11.0	26.8
β-Lactalbumin	1.2	3.4	19.1	22.8
Pepsinogen	0.8	—	27.2	31.5

Tanford, C. *et al.* (1967) *J. Am. Chem. Soc.,* **89**, 729.

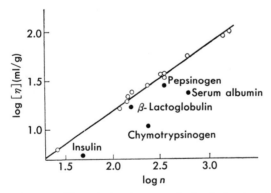

Fig. 5-1. The relation between the intrinsic viscosity ($[\eta]$) and the number of residues (n) for intact proteins (●) and reduced proteins (○) in 6 M guanidine hydrochloride.

restrict the dimensions of the protein molecule. This restriction depends upon the positions of the disulphide bonds. For hen egg-white lysozyme, the intrinsic viscosity is 2.7 ml/g under native conditions and 6.5 ml/g in 6 M guanidine hydrochloride. This indicates that the molecular volume is increased only by a factor of about two on denaturation in 6 M guanidine hydrochloride. Under such conditions, the residues are located close to each other and contacts among side chains and between the side chains and the main chain will occur more frequently than in the reduced protein. However, as described in Section 4-3, the ionization behaviour of proteins in 6 M guanidine hydrochloride can be explained by assuming a random coil without non-covalent interactions. It has also been shown that the residue optical rotation of a protein in 6 M guanidine hydrochloride can be expressed as the sum of the intrinsic residue rotation of each amino acid residue (Tanford *et al.*, 1967).

As described above, measurements of various physicochemical properties show that proteins behave as a random coil in 6 M guanidine hydrochloride. Thus studies of protein denaturation by guanidine hydrochloride are very helpful for elucidating the equilibrium and kinetics of the transition between the native and completely random states of a protein.

Another important denatured state is that termed the "molten globule" (see for instance, Kuwajima, 1989; Ptitsyn, 1987). This term was first applied by Ohgushi and Wada (1983). This state is realized for proteins in media at acidic or alkaline pH in the presence of salts at moderate concentrations. A protein in the molten globule state shows a CD spectrum essentially identical to that of the native protein in the far ultraviolet region, but similar to that of the completely unfolded protein in the aromatic region. The protein molecule in the molten globule state is as compact as the protein in the native state. Thus the molten globule state is defined as the state of a protein with native secondary structure and compactness but with fluctuating tertiary structure. Recent NMR studies (Baum *et al.*, 1989; Hughson *et al.*, 1990; Jeng *et al.*, 1990) have shown that in cytochrome *c*, α-lactalbumin and apomyoglobin in the molten globule state, secondary structures are, at least in part, formed in the same segments as those existing in the native proteins. The molten globule state has low cooperativity in unfolding transitions and is in rapid equilibrium with the denatured state. A concentrated solution of guanidine hydrochloride or urea is a good solvent for denatured proteins. The solvent-solute interactions are thus favourable and the solute molecule assumes an extended random conformation. In contrast, water is a poor solvent for denatured proteins and the solvent-solute interactions are unfavourable. The solute molecule thus tends to assume a compact conformation, and if there is a sequence with high propensity for α-helix formation, the segment will form an α-helix in water. The molten globule state also appears as an early intermediate in the refolding process. This will be described in Chapter 7.

5-2. TWO-STATE TRANSITION

When a physical or chemical property (y) reflecting protein conformation is followed as the concentration of a denaturant or temperature increases, the value of y changes rapidly at a given concentration of the denaturant or temperature and reaches the value of y characteristic of the denatured protein (Fig. 5-2). This indicates that once a conformational change begins to occur, the whole molecule is unfolded completely, that only the native protein molecule and completely unfolded protein molecule exist, and that no partially unfolded molecule exists in the denaturation process between y_N and y_D. Such a transition is referred to as a "two-state transition". For this transition, the following equilibrium holds between the native protein (N) and unfolded protein (D).

$$N \rightleftharpoons D \tag{5.2}$$

Denaturation of most proteins can be approximated by the two-state transition, but some proteins proceed through two or more steps in the course of denaturation. In the latter case, one or more unfolding intermediates exist. Whether or not denaturation

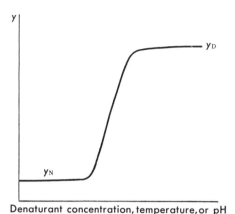

Denaturant concentration, temperature, or pH

Fig. 5-2. A typical protein unfolding curve. y represents a quantity which reflects any of the physical and chemical properties of the protein molecule. y_N and y_D express y for native and unfolded proteins, respectively.

proceeds through a two-state transition depends upon the protein and the denaturation conditions. Denaturation of a protein that consists of two or more domains will proceed through two or more steps when the domains have different stabilities towards the denaturant, even if the denaturation of each domain can be approximated by a two-state transition. When the denaturation is not approximated by a two-state transition, it is not easy to analyse the denaturation curve. However, if there is partial unfolding in the course of denaturation, this will yield information on the unfolding process.

The following methods are used to test whether the denaturation of a protein under study is approximated by a two-state transition. (1) The denaturation is examined by various methods such as CD in the far-ultraviolet region, which reflects the polypeptide backbone conformation, CD in the aromatic absorption region, which reflects the states of aromatic residues, and fluorescence, which reflects the states of aromatic resides, and chemical modification of a given side chain. If all these methods give the same denaturation curve, then the denaturation can be expressed by a two-state transition. (2) Assuming a two-state transition, $N \rightleftharpoons D$, the equilibrium constants (K_D) are determined at various temperatures, and the enthalpy change (ΔH_D) is estimated using the van't Hoff equation

$$\Delta H_D = -R\frac{d\ln K_D}{d(1/T)}. \tag{5.3}$$

If the enthalpy change thus obtained is the same as that determined directly by calorimetry, then the denaturation can be approximated by a two-state transition.

When the observed value used to examine unfolding is expressed by y, the following equation is obtained for a two-state transition

$$y = f_N y_N + f_D y_D, \tag{5.4}$$

where y_N and y_D are the values of y for the N state and D state, respectively, and f_N and f_D are the fractions of proteins in the N state and D state, respectively. f_D is expressed by

$$f_D = 1 - f_N = \frac{y - y_N}{y_D - y_N}. \tag{5.5}$$

The equilibrium constant (K_D) for the reaction N$\rightleftharpoons$D is expressed by

$$K_D = e^{-\Delta G_D/RT} = \frac{f_D}{1 - f_D} = \frac{y - y_N}{y_D - y}, \tag{5.6}$$

where ΔG_D is the free energy change of unfolding.

It is necessary to examine the reversibility of unfolding when the above calculation is made. A protein solution at a high concentration of denaturant is diluted gradually with solvent containing no denaturant and the refolding curve is measured. If the refolding curve thus obtained is the same as the unfolding curve obtained by increasing the denaturant concentration, then the protein is shown to be reversibly unfolded.

If there are stable intermediates (X_1, X_2, $\cdots$) in addition to proteins in the N and D states, y at a given point in the unfolding curve is given by

$$y = f_N y_N + f_D y_D + \Sigma f_{Xi} y_{Xi}, \tag{5.7}$$

where y_{Xi} is the value of y for X_i and f_N, f_D and f_{Xi} are the fractions of proteins in N, D and X_i states, respectively.

f_N is given by

$$f_N = 1 - f_D - \Sigma f_{Xi}. \tag{5.8}$$

The apparent fraction of unfolded protein f_{app} is given by

$$f_{app} = \frac{y - y_N}{y_D - y_N} = \Sigma f_{Xi} \alpha_i + f_D, \tag{5.9}$$

where

$$\alpha_i = \frac{y_{Xi} - y_N}{y_D - y_N}. \tag{5.10}$$

Let the apparent equilibrium constant K_{app} be

$$K_{app} = \frac{f_{app}}{1 - f_{app}}, \tag{5.11}$$

where $(1 - f_{app})$ is the apparent fraction of unfolded protein.

The true equilibrium constants K_i and K_D are defined by

$$K_i = \frac{f_{Xi}}{f_N} \tag{5.12}$$

$$K_D = \frac{f_D}{f_N}. \tag{5.13}$$

From Eqs. (5.8), (5.9) and (5.11), the following equations are obtained

$$\frac{f_{obs}}{f_N} = K_D(1 + \Sigma \alpha_i K_i / K_D) \tag{5.14}$$

$$\frac{1 - f_{app}}{f_N} = 1 + \Sigma(1 - \alpha_i) K_i \tag{5.15}$$

$$K_{app} = K_D \frac{1 + \Sigma \alpha_i K_i / K_D}{1 + \Sigma(1 - \alpha_i) K_i} \tag{5.16}$$

When only a single partially unfolded intermediate (I) exists in an unfolding process,

$$N \underset{\longleftarrow}{\overset{K_1}{\rightleftharpoons}} I \underset{\longleftarrow}{\overset{K_2}{\rightleftharpoons}} D, \tag{5.17}$$

the apparent equilibrium constant is given by

$$K_{app} = \frac{K_D + \alpha K_1}{1 + (1 - \alpha) K_1} = \frac{K_1 K_2 + \alpha K_1}{1 + (1 - \alpha) K_1}. \tag{5.18}$$

5-3. EFFECTS OF DENATURING AGENTS

A) Urea and Guanidine Hydrochloride

Urea and guanidine hydrochloride are typical denaturants which have been used frequently. However, aqueous solutions of urea yield cyanate (N=C-OH) on standing, which carbamylates various amino acid side chains of proteins. The equilibrium concentration of cyanate in 8 M urea is approximately 0.02 M. Therefore, recrystallization or deionization of urea solutions should be carried out just before experiments. Since guanidine hydrochloride has no such disadvantage and the properties of proteins in 6 M guanidine hydrochloride have been well established (see Section 5-1), denaturation by guanidine hydrochloride has been studied more extensively. The denaturing power of guanidinium salts increases in the order Cl < Br < I < SCN (Castellino & Barker, 1968).

Using the procedure described in Section 1-3, we can determine the free energy changes (Δg_t) of transfer of amino acid side chains and peptides from water to an aqueous denaturant solution. The value of Δg_t gives a measure of the affinity of a side chain or a peptide group for a denaturant, and is thus useful for clarifying the mechanism of protein denaturation.

The free energy changes of transfer of amino acid side chains and peptides from water to aqueous solutions of urea and guanidine hydrochloride are given in Tables 5-2 and 5-3, respectively. When the Δg_t values are compared at the same concentration,

TABLE 5-2

Changes in Free Energy (Δg_t) of Transfer of Amino Acid Side Chains and Peptides from Water to Urea Solutions (25°C)

Side chain or peptide	Δg_t (cal/mol)			
	2 M	4 M	6 M	8 M
Ala	0	+15	+10	+10
Leu	−110	−155	−225	−295
Phe	−180	−330	−470	−600
Trp	−270	−505	−730	−920
Met	−115	−225	−325	−415
Thr	−40	−60	−90	−115
Tyr	−225	−395	−580	−735
His	−100	−160	−205	−255
Asn	−135	−225	−330	−430
Gln	−80	−130	−190	−230
Peptide				
$(Gly)_2 - (Gly)$	−10	−20	−35	−60
$(Gly)_3 - (Gly)_2$	−145	−205	−305	−310
Carbobenzoxydiglycine− carbobenzoxyglycine	−20	−45	−5	+5
Acetyltetraglycine- ethylester − butane	−36	−72	−109	−145

Nozaki, Y. & Tanford, C. (1963) *J. Biol. Chem.*, *238*, 4074.

the negative value of Δg_t for any side chain in guanidine hydrochloride is two to three times greater than that in urea. This indicates that amino acid side chains or a peptide group have greater affinity for guanidine hydrochloride than for urea, and suggests that the denaturing action of guanidine hydrochloride is two to three times stronger than that of urea. In fact, the concentration of guanidine hydrochloride required to denature a protein is generally about half the urea concentration required to denature the same protein. The Δg_t values for urea and guanidine hydrochloride of hydrophobic side chains as well as peptide groups are negative. This indicates that urea and guanidine hydrochloride also have affinity for hydrophobic side chains, particularly those of Phe, Tyr and Trp. When the number of guanidine hydrochloride molecules bound to 13 different proteins is plotted against the sum of the number of aromatic residues and half the total number of amino acid residues, a straight line with a slope of 1 is obtained.

TABLE 5-3

Changes in Free Energy (Δg_t) of Transfer of Amino Acid Side Chains and Peptides from Water to Guanidine Hydrochloride Solutions (25.1°C)

Side chain or peptide	Δg_t (cal/mol)			
	1 M	2 M	4 M	6 M
Ala	−10	−20	−30	−45
Leu	−150	−210	−355	−480
Phe	−215	−335	−580	−775
Trp	−400	−630	−980	−1,235
Met	−150	−245	−400	−535
Thr	−65	−90	−120	−125
Tyr	−235	−385	−605	−770
His	−180	−285	−385	−420
Asn	−200	−320	−490	−645
Gln	−135	−215	−315	−360
Peptide				
(Gly)$_2$−(Gly)	−85	−120	−175	−220
(Gly)$_3$−(Gly)$_2$	−220	−305	−395	−420
(acetyltetraglycine-ethylester − ethylacetate)/4	−85	−135	−205	−245

Nozaki, Y. & Tanford, Y. (1970) *J. Biol. Chem., 245*, 1648.

This suggests that guanidine hydrochloride binds mainly to aromatic side chains and peptide groups.

The changes in free energy of transfer of amino acid side chains from water to 8 M urea and to 6 M guanidine hydrochloride increase linearly with the increase in accessible surface area of the side chains with slopes of 7.1 and 8.3 cal/mol/Å^2, respectively. As described in Section 3-2, the slope of the plot of the changes in free energy of transfer of side chains from water to organic solvent against the accessible surface area was found to be 24 cal/mol/Å^2. Thus 8 M urea or 6 M guanidine hydrochloride decreases the unfavourable hydrophobic interactions of side chains with water to about one third.

B) Organic Solvents

The free energy changes (Δg_t) for transfer of amino acid residues and peptides from water to ethylene glycol and dioxane at

TABLE 5-4

Changes in Free Energy (Δg_t) of Transfer of Amino Acid Side Chains and Peptides from Water to Aqueous Ethylene Glycol Solutions (25°C)

Side chain or peptide	Δg_t (cal/mol)		
	30%	60%	90%
Ala	+30	+5	−20
Leu	−130	−355	−815
Phe	−215	−665	−1,115
Trp	−515	−1,310	−2,345
Met	−65	−290	−670
Thr	+35	+55	+30
Tyr	−305	−790	−1,410
His	−115	−295	−485
Asn	−80	−450	−830
Gln	+10	+40	+60
Peptide			
$(Gly)_2 - (Gly)$	+75	+170	+280
$(Gly)_3 - (Gly)_2$	+65	+115	+240

Nozaki, Y. & Tanford, C. (1965) *J. Biol. Chem.*, *240*, 3568.

TABLE 5-5

Changes in Free Energy (Δg_t) of Transfer of Amino Acid Side Chains and Peptides from Water to Aqueous Dioxane Solutions (25.1°C)

Side chain or peptide	Δg_t (cal/mol)						
	20% (2.3 M)	30% (3.5 M)	40% (4.73 M)	60% (7.0 M)	80% (9.7 M)	90% (10.5 M)	100% (11.7 M)
Leu	−165		−415	−845	−1,190	−1,310	−1,400
Phe	−375	−600	−845	−1,430	−1,940	−2,160	−2,320
Tyr	−465		−1,055	−1,560	−1,955	−2,125	−2,250
Trp	−765	−1,175	−1,610	−2,495	−3,145	−3,345	−3,480
His	−115	−150	−205	−315			−480
Asn	−25		−35	−225			
Gln	−10		−10	−230			
Peptide							
$(Gly)_2 - (Gly)$	120		330	585			
$(Gly)_3 - (Gly)_2$	55		90	205			

Nozaki, Y. & Tanford, C. (1971) *J. Biol. Chem.*, *246*, 2211.

various concentrations are shown in Tables 5-4 and 5-5, respectively. The Δg_t values are negative for hydrophobic residues but positive for peptide groups. It is often observed that the helical content of proteins increases in the presence of organic solvents.

The dielectric constants of aqueous organic solvents are smaller than that of pure water and thus the presence of an organic solvent increases the electrostatic interactions between charged groups. The denaturing action of ethylene glycol is very small. Ribonuclease A and hen egg-white lysozyme are not denatured even in 100% ethylene glycol.

C) Salts

Salts act on protein conformations in two ways: (1) the Debye-Hückel screening effect of ions and (2) binding of ions to proteins. At low concentrations, the Debye-Hückel effect is dependent only on the ionic strength of the medium and not on ion species. At moderate concentrations, anions or cations bind to positively or negatively charged groups of proteins and stabilize the protein conformation by reducing the repulsion between charged groups of the same sign. Cytochrome c, β-lactamase, apomyoglobin and interferon are unfolded maximally at pH 2 under conditions of low ionic strength. On increasing the concentration of salts at pH 2, these proteins fold to a molten globule state (Goto et al., 1990a, b; Goto & Fink, 1990). This effect is dependent on ion species, and the order of anion effectiveness is ferricyanide > ferrocyanide > sulphate > trichloroacetate > thiocyanate > perchlorate > iodide > nitrate > trifluoroacetate > bromide > chloride. This series is similar to the Hofmeister (lyotropic) series found by Hofmeister as early as 1888 in his study on the solubility of proteins, and also resembles the electroselectivity series of anions toward the anion-exchange resins (Gjerde et al., 1980; Washabaugh & Collins, 1986).

Salts at high concentration denature proteins. The denaturing power increases in the order $SO_4^{2-} < CH_3COO^- < Cl^- < Br^- < ClO_4^- < CNS^-$ for anions and $(CH_3)_4N^+$, NH_4^+, K^+, $Na^+ < Li^+ < Ca^+$ for cations (von Hippel & Schleich, 1969). The same series was also obtained for the denaturation of DNA (Hamaguchi & Geiduschek, 1962): Cl^-, $Br^- < CH_3COO^- < I^- < ClO_4^- < CNS^-$ for anions and $(CH_3)_4N^+ < K^+ < Na^+ < Li^+$ for cations. These series were called the "chaotropic"* series (Hamaguchi & Geiduschek, 1962). Although the mechanism of the denaturing action of concentrated salts is not well known, the stronger the affinity of the ion to proteins, the stronger the denaturing power.

Table 5-6 shows the changes in free energy of transfer of the side chains of Ala, Val, Leu and Phe and a peptide group from water to aqueous solutions of various salts. The data in this table suggest that the hydrophobic interactions become stronger and the hydrogen bonds between peptide groups become weaker in concentrated salt solutions.

The sulphate ions of ammonium sulphate and sodium sulphate stabilize proteins. The V_L and C_L domains of the immunoglobulin light chain can be isolated as fragments by limited proteolysis, and these fragments are completely unfolded in 1.8 and 1.5 M guanidine hydrochloride, respectively. As ammonium sulphate is added to these solutions of the denatured proteins, the original native conformations are recovered (Fig. 5-3). The thermal transition temperatures are shifted to higher temperatures in the presence of sulphate ions, and kinetic studies have shown that ammonium sulphate stabilizes the V_L and C_L fragments mainly by decreasing the unfolding rate (Goto et al., 1988).

Recently the structures of ribonuclease A with a sulphate ion bound at the active site and the sulphate-free protein have been compared by X-ray crystallographic analysis (Campbell & Petsko,

TABLE 5-6

Changes in Free Energy (Δg_t) of Transfer of the Side Chains of Ala, Val, Leu, and Phe and Peptide from Water to Various Aqueous Salt Solutions (25°C)

	1 M Na$_2$SO$_4$	2 M NaCl	2 M NaBr	2 M NaClO$_4$	2 M NaSCN	2 M LiCl	2 M CaCl$_2$
Ala	+130	+90	—	—	—	+85	+220
Val	+370	+31	+320	+190	+250	+250	+800
Leu	+410	+335	+260	+275	+190	+320	+635
Phe	+425	+325	+250	+65	—	+295	+220
Peptide	−15	−60	−95	−180	−170	−80	−190

Nandi, P.K. & Robinson, D.R. (1972) J. Am. Chem. Soc., 94, 1299, 1308.

*Tending to disorder

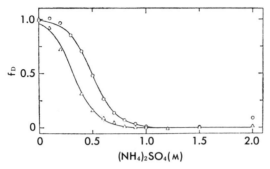

Fig. 5-3. Refolding transitions by ammonium sulphate of the C_L fragment in 1.5 M guanidine hydrochloride (circles) and of the V_L fragment in 1.8 M guanidine hydrochloride (triangles) at pH 7.5 and 25°C. The ordinate indicates the fraction of the unfolded molecule. (Goto, Y. *et al.* (1988) *Biochemistry, 27*, 1670)

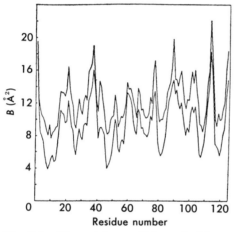

Fig. 5-4. Thermal factors (B factors) of the main chain of ribonuclease A plotted against the residue number. Lower and upper curves represent the thermal factors for sulphate-bound and sulphate-free proteins, respectively. (Campbell, R.I. & Petsko, G.A. (1987) *Biochemistry, 26*, 8579)

1987). The crystallographic B factors (see Section 6-2) of sulphate-free ribonuclease A are greater than those of the sulphate-bound protein, indicating an overall loosening of the protein molecule by removal of the sulphate ion (Fig. 5-4). It is not certain whether this finding is related to the stabilizing action of sulphate ion for proteins in general. Application of the data in Table 5-6 to the interpretation of protein stability will be given in Section 5-4.

Ammonium sulphate has long been used to purify proteins because its solubility in water is extremely high and thus is suited for salting-out of proteins. The stabilizing action of sulphate ions on proteins is thus also a fortuitous property for this purpose.

D) pH

Proteins are generally stable at neutral pH and are denatured at extreme pH values. In the native protein molecule, an ionizable group is sometimes found to have a pK_a value quite different from its normal one (Table 1-2) (see Section 4-3), but such abnormal pK_a values revert to normal in the unfolded state. When the pK_a value in the native state changes on unfolding, the affinity of a proton for the ionizable group in the unfolded state differs from that in the native state, and this is responsible for the unfolding of proteins at acidic or alkaline pH values. The following equation has been derived by Tanford (1970)

$$K_{\mathrm{D}} = K_0 \frac{\prod_{j=1}^{n} \{(\mathrm{H}^+) + K_{\mathrm{a},j,\mathrm{D}}\}}{\prod_{j=1}^{n} \{(\mathrm{H}^+) + K_{\mathrm{a},j,\mathrm{N}}\}}, \qquad (5.19)$$

where K_{D} is the equilibrium constant for the two-state transition of $\mathrm{N} \rightleftharpoons \mathrm{D}$, (H^+) is the activity of hydrogen ions, $K_{\mathrm{a},j,\mathrm{N}}$ and $K_{\mathrm{a},j,\mathrm{D}}$ are the ionization constants (K_{a}) of ionizable group j in the N and D states, respectively, and K_0 is the equilibrium constant for the reaction $\mathrm{N} \rightleftharpoons \mathrm{D}$ when $K_{\mathrm{a},j} \ll (\mathrm{H}^+)$.

Let us explain the unfolding by acid and alkali of the C_{L} fragment of an immunoglobulin light chain. Figure 5-5 shows the changes with pH in the ellipticity at 218 nm for the intact and reduced C_{L} fragments. The C_{L} fragment has only one intrachain disulphide bond buried in the interior of the protein molecule. Reduction of the disulphide bond does not cause any conformational change in the molecule (see Section 5-5). The C_{L} fragment has two His residues at positions 189 and 198 whose pK_a values were determined to be 7.5 and 4.6, respectively, by titration fol-

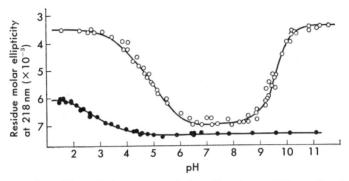

Fig. 5-5. The pH dependence of the ellipticity at 218 nm for the intact C_L fragment (●) and reduced C_L fragment (○). The solid lines represent the theoretical curves constructed using Eqs. (5.19) and (5.20). (Ashikari, Y. *et al.* (1985) *J. Biochem.*, 97, 517)

lowed by NMR spectroscopy (Shimizu *et al.*, 1980). These pK_a values were not changed on reduction of the disulphide bond (Ashikari *et al.*, 1985). The pK_a value of His 189 is slightly higher, and that of His 198 is abnormally lower than the normal pK_a value of the His residue. If the conformations of the intact C_L and reduced C_L fragments are the same, the number and the pK_a values of ionizable groups which participate in the acid unfolding should be the same, and the difference in the unfolding behaviour between the two proteins should be explained only in terms of the presence or absence of the disulphide bond. The equilibria of unfolding by acid of the intact C_L and reduced C_L fragments were found to be explained well by assuming that ionizable groups with $pK_{a,1,N}=7.0$, $pK_{a,2,N}=4.8$ and $pK_{a,3,N}=2.6$ of both native proteins are changed to $pK_{a,1,D}=pK_{a,2,D}=6.5$ and $pK_{a,3,D}=4.0$ on unfolding by acid (see Eq. 5.19). The ionizable groups with $pK_{a,1}$ and $pK_{a,2}$ may correspond to His 189 and His 198, respectively, and these pK_a values revert to the normal pK_a value of 6.5 for His on unfolding. The ionizable group with $pK_{a,3}$ may be carboxyl and the pK_a value changes to the normal pK_a value of 4.0 for the carboxyl on unfolding. The difference in stability between the two proteins in the acidic region was explained by assuming that the stability of the intact C_L fragment is 100 times greater than that of the reduced C_L fragment.

In the alkaline pH region, the reduced C_L fragment was unfolded above pH 8.5, while the intact C_L fragment was not unfolded until pH 11.6. In the alkaline pH region, the two SH groups of the reduced C_L fragment ionize, while these groups are not present in the intact C_L fragment. The unfolding by alkali of the reduced C_L fragment is explained by assuming that the reduced C_L fragment can no longer adopt the folded conformation when either or both of the SH groups ionize. On this assumption, the following equation is derived

$$K_D = K_0' \left(1 + \frac{K_a}{(H^+)}\right)^2, \tag{5.20}$$

where K_0' is the equilibrium constant for the reaction $N \rightleftharpoons D$ at neutral pH and K_a is the ionization constant of the SH group in the unfolded molecule. This equation can also be derived from Eq. (5.19) by setting $n = 2$, $K_{a,1,N} = K_{a,2,N} = 0$, $K_{a,1,D} = K_{a,2,D} = K_a$ and $K_0 = K_0'$.

The unfolding equilibrium by alkali of the reduced C_L fragment was found to be explained well by placing $K_a = 9.3$ in Eq. (5.20).

As described above, cytochrome c, β-lactamase and apomyoglobin are maximally unfolded at pH 2 under conditions of low ionic strength. However, a further decrease in pH by increasing the concentration of HCl refolds these proteins to the molten globule state. This is caused by binding of chloride ions to positively charged groups (Goto *et al.*, 1990a, b).

E) Heat

Studies on thermal unfolding are important because thermodynamic parameters of unfolding are obtained by analysing the reversible thermal transitions. The heat capacity increases on protein unfolding. Owing to the difference in heat capacity between native and denatured proteins, the enthalpy change ($\Delta C_p =$

$T(\partial \Delta H / \partial T)$p) and entropy change $(\Delta C_p = T(\partial \Delta S / \partial T)$p) on unfolding change greatly with temperature. The changes in free energy, enthalpy and entropy with temperature for ribonuclease A are shown in Figs. 5-6 and 5-7.

In the temperature range where the heat capacity change does not vary with temperature, the temperature dependence of ΔH and that of ΔS are expressed by

$$\Delta H(T) = \Delta H(T_m) + \Delta C_p(T - T_m) \qquad (5.21)$$

$$\Delta S(T) = \Delta S(T_m) + \Delta C_p \ln(T - T_m) \qquad (5.22)$$

where T_m is the temperature at the midpoint of the thermal unfolding curve. The change in free energy is zero at $T = T_m$.

Privalov (1979) showed that plots of the changes in enthalpy of unfolding per gram of protein against temperature intersect at a common temperature of 110°C for most proteins, that the slopes of these plots, ΔC_p, are proportional to the content of hydrophobic residues: the plots of the changes in entropy of unfolding per gram of protein also intersect at around 110°C.

As described in Section 4-2, Baldwin (1986) attempted to explain the hydrophobic interactions of proteins on the basis of the

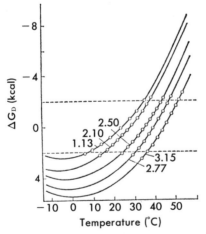

Fig. 5-6. Changes in free energy on thermal unfolding of ribonuclease A at various pH values. (Brandts, J.F. & Hunt, L. (1967) *J. Am. Chem. Soc., 89*, 4826)

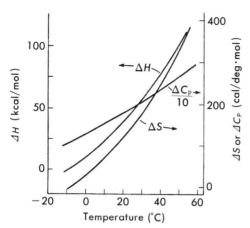

Fig. 5-7. Changes in enthalpy, entropy and heat capacity on thermal unfolding of ribonuclease A. (Brandts, J.F. & Hunt, L. (1967) *J. Am. Chem. Soc., 89,* 4826)

thermodynamic properties of the dissolution of hydrocarbons in water. He was able to explain Privalov's finding that the plots of the changes in entropy of unfolding intersect at 110°C, but not the finding that the plots of changes in enthalpy of unfolding intersect at 110°C. Baldwin also estimated the hydrophobic contribution to the stability of hen egg-white lysozyme by analysing the thermal unfolding curve in terms of the thermodynamic properties of hydrocarbons. The hydrophobic contributions to $\Delta H°$ and $\Delta S°$ were estimated using Eqs. (4.12) and (4.13), respectively. The change in free energy ($\Delta G°_{hyd}$) due to hydrophobic interactions is expressed by

$$\Delta G°_{hyd} = \Delta C_{phyd}(T - T_h) + \Delta C_{phyd} T \ln(T_s/T). \qquad (5.23)$$

In these calculations, T_s and T_h obtained with hydrocarbons ($T_h =$ 295K and $T_s = 380$K) were used. The heat capacity change due only to hydrophobic interactions (ΔC_{phyd}) is unknown. However, if the heat capacity change on unfolding (1,650 cal/mol/deg) is assumed to be due to hydrophobic interactions alone, then the thermodynamic parameters for unfolding can be estimated (Table 5-7). In this table, $\Delta H°_{res}$, $\Delta S°_{res}$ and $\Delta G°_{res}$ represent the difference between the observed and calculated values. The values of $\Delta H°_{res}$

TABLE 5-7
Thermodynamic Parameters for Unfolding of Hen Egg-white Lysozyme

Temperature (°C)	$\Delta H°$ (cal/mol)			$\Delta S°$ (cal/mol/deg)			$\Delta G°$ (cal/mol)		
	Obs	Hyd	Res	Obs	Hyd	Res	Obs	Hyd	Res
10	32.8	−18.6	51.4	59.0	−484	543	16.1	118	102
25	56.4	4.8	51.6	140	−403	544	14.5	125	110
60	112	59.3	52.8	315	−230	544	6.5	136	130
100	175	121.7	53.3	494	−53.5	548	9.9	142	131
		Average	52.3			545			

Hyd, contributions of hydrophobic interactions; Res, contributions other than hydrophobic interactions; Obs, observed.
Baldwin, R.L. (1986) *Proc. Natl. Acad. Sci. U.S.A., 83*, 8069.

and $\Delta S°_{res}$ are independent of temperature and are found to be 52 kcal/mol and 545 cal/mol/deg, respectively. This value of $\Delta S°_{res}$ agrees well with the value (547 cal/mol/deg) estimated from the intersection at $T = T_s$ in the plot of $\Delta S°_{res}$ against $\ln(T/T_s)$.

Recently, there have been many studies on the thermal stability of mutant proteins in which a particular amino acid residue has been replaced by genetic engineering (see Section 5-5). The difference between the change in free energy of unfolding ($\Delta\Delta G$) at T_m for a mutant protein and the reference protein is expressed by the following equation (Becktel & Schellman, 1987)

$$\Delta\Delta G = \Delta T \Delta S, \qquad (5.24)$$

where ΔT is the difference in T_m and ΔS is the entropy of unfolding of the reference protein at T_m.

5-4. STABILITY OF PROTEINS IN AQUEOUS SOLUTIONS

One of the most interesting questions in studies of protein unfolding is to know by what difference in free energy the native regular conformation of a protein is stabilized compared with its unfolded protein under physiological conditions. The following five experimental methods are employed to estimate this difference in free energy.

1) The unfolding curve of a protein is determined using guanidine hydrochloride or urea as a denaturant. Assuming a two-state transition, the change in free energy (ΔG_D) at each denaturant concentration is estimated using Eq. (5.6). Tanford (1970) derived the following equation assuming that the difference in the number of binding sites of guanidine hydrochloride between the native and unfolded proteins affects the equilibrium constant for the reaction $N \rightleftharpoons D$*

$$G_D = \Delta G_D^{H_2O} - \Delta n \ RT \ \ln \ (1 + ka_{\pm}), \qquad (5.25)$$

where $\Delta G_D^{H_2O}$ is the free energy difference between the native and unfolded states in aqueous solution, Δn is the difference in the number of binding sites between the unfolded and native states, k is the average binding constant of the sites, and $a_{\pm}$ is the mean ion activity of guanidine hydrochloride or urea. As the value of k, 0.6 M^{-1} and 0.1 M^{-1} are used for guanidine hydrochloride and urea, respectively (Pace, 1986; Pace *et al.*, 1989). The mean ion activity of guanidine hydrochloride is related to molar concentration (M) by the equation

$$a_{\pm} = 0.676(M) - 0.1468(M)^2 + 0.02475(M)^3 - 0.00132(M)^4. \qquad (5.26)$$

As can be seen from Eq. (5.25), the plot of ΔG_D against $\ln(1 + ka_{\pm})$ gives a straight line and the values of $\Delta G_D^{H_2O}$ and Δn are obtained from extrapolation of the straight line to $a_{\pm} = 0$ and the slope, respectively.

2) When the values of ΔG_D at various concentrations of guanidine hydrochloride or urea are plotted against the denaturant concentrations, a straight line is obtained which is expressed by the following equation. The value of $\Delta G_D^{H_2O}$ can be estimated by extrap-

---

*When the pK_a value of an ionizable group is changed on unfolding, the binding ability of a proton is changed even though the number of proton binding sites remains unchanged. Equation (5.19) was derived on this assumption. For unfolding by guanidine hydrochloride or urea, the number of binding sites of the denaturant is changed on unfolding, although the binding constant of the denaturant remains unchanged. On this assumption, Eq. (5.25) was derived. Thus the basic concepts for derivation of Eqs. (5.19) and (5.25) are the same.

olation of the straight line to zero concentration of denaturant.

$$\Delta G_D = \Delta G_D^{H_2O} - m(\text{denaturant}), \tag{5.27}$$

where (denaturant) is the concentration of denaturant and m is a measure of the dependence of ΔG_D on denaturant concentration, being proportional to the difference in accessible surface area between the native and unfolded proteins (Schellman, 1978). Since $\Delta G_D = 0$ at the midpoint of the denaturation curve

$$(\text{denaturant})_{1/2} = \Delta G_D^{H_2O}/m. \tag{5.28}$$

The values of $\Delta G_D^{H_2O}$ for a protein obtained using Eqs. (5.25) and (5.27) are almost the same, although that from Eq. (5.25) is always somewhat larger. It is unknown at present which method is more appropriate for estimating the value of $\Delta G_D^{H_2O}$ (Pace, 1986).

The values of $\Delta G_D^{H_2O}$ obtained using guanidine hydrochloride and urea as denaturants are almost the same for most proteins. However, it has recently been reported for rat intestinal fatty acid-binding protein that the value of $\Delta G_D^{H_2O}$ (5.2 kcal/mol) obtained from guanidine hydrochloride denaturation data differs greatly from the value (10.0 kcal/mol) obtained using urea denaturation data (Ropson *et al.*, 1990). The reason for this is not clear.

3) Let us consider the following cycle (Tanford, 1970)

$$
\begin{array}{ccc}
\text{Native protein} & \xrightarrow{\Delta G_D^{H_2O}} & \text{Unfolded protein} \\
\text{(aqueous solution)} & & \text{(aqueous solution)} \\
\Delta G_{t,N} \downarrow & & \downarrow \Delta G_{t,D} \\
\text{Native protein} & \xrightarrow{\Delta G_D} & \text{Unfolded protein} \\
\text{(denaturant solution)} & & \text{(denaturant solution).}
\end{array}
\tag{5.29}
$$

The following relation holds for this cycle

$$\Delta G_D - \Delta G_D^{H_2O} = \Delta G_{t,D} - \Delta G_{t,N}, \tag{5.30}$$

where $\Delta G_{t,N}$ and $\Delta G_{t,D}$ are the changes in free energy of transfer of native and unfolded proteins, respectively, from aqueous solution to a denaturant solution. The difference between $\Delta G_{t,D}$ and $\Delta G_{t,N}$ is related only to the groups which are exposed on unfolding and not to the groups which are initially exposed in the native state.

When the fraction α_i of the total number of group i (n_i^0) is buried in the interior of the protein molecule, we may write

$$\Delta G_{t,D} - \Delta G_{t,N} = \sum_i \alpha_i n_i^0 \Delta g_{t,i}, \tag{5.31}$$

where $\Delta g_{t,i}$ is the change in free energy of transfer of group i from aqueous solution to a denaturant solution (see Tables 5-2 to 5-6). When the average value of α for various groups is expressed by $\bar{\alpha}$, the following equation is obtained

$$\Delta G_D = \Delta G_D^{H_2O} + \bar{\alpha} \sum n_i^0 \, \Delta g_{t,i}. \tag{5.32}$$

A value of 0.35 is often used for $\bar{\alpha}$. Since $\Delta G_D = 0$ at the midpoint of an unfolding curve, $\Delta G_D^{H_2O}$ can be estimated from the equation

$$\Delta G_D^{H_2O} = -\bar{\alpha} \sum n_i^0 \, \Delta g_{t,i}. \tag{5.33}$$

The $\Delta G_D^{H_2O}$ value thus obtained is larger than that estimated using Eq. (5.25). The values of $\Delta G_D^{H_2O}$ of various proteins obtained using the above three methods are shown in Table 5-8. As described at the beginning of Chapter 4, the change in free energy for the reaction $N \rightleftharpoons D$ is very small.

Application of Eq. (5.33) to protein denaturation by concentrated salt solutions will be described. As shown in Table 5-6, the Δg_t values were not determined for all the amino acid side chains. Assuming that the value of Δg_t for Tyr is the same as that for Phe, and the Δg_t values of Ile, Leu, Pro and Met are the same as that for Val, and using the values of Ala and the peptide group given in Table 5-6, Nandi and Robinson (1972b) estimated the difference in the free energy of ribonuclease A between water and a salt solution using Eq. (5.33). In this estimate, the contributions of the ionizable and hydrophilic side chains were neglected, because these side chains are largely exposed to solvent in both the native and unfolded states. The results are shown in Table 5-9. The values of $0.35 \sum n_i^0 \Delta g_{t,i}$ for 2 M NaSCN and 2 M CaCl$_2$ are negative, indicating that the stability of ribonuclease A is decreased in these solvents; indeed the melting temperatures of thermal unfolding are decreased. On the other hand, the values of $0.35 \sum n_i^0 \Delta g_{t,i}$ for 1 M Na$_2$SO$_4$ and 2 M NaCl are positive, indicating that the stability of the protein is increased in these solvents, and the melting tempera-

TABLE 5-8
Changes in Free Energy ($\Delta G_D^{H_2O}$) of Unfolding of Proteins in Water (25°C)

Protein	$\Delta G_D^{H_2O}$ (kcal/mol)	Protein	$\Delta G_D^{H_2O}$ (kcal/mol)
C_L fragment of immunoglobulin	5.7	Barnase	9.4
Nuclease	6.1	Ribonuclease T1	10.2
Ribonuclease A	9.2	T4 phage lysozyme	12.8
Hen lysozyme	9.3	Bovine pancreatic trypsin inhibitor	14.3

TABLE 5-9
Unfolding of Ribonuclease A by Various Salts   (cal/mol)

	1 M Na$_2$SO$_4$	2 M NaCl	2 M NaSCN	2 M CaCl$_2$
Non-polar Side chains	+13,900	+11,100	+3,000	+22,500
Peptides	−1,900	−7,400	−20,900	−23,400
$0.35\Sigma n_i^0 \Delta g_{t,i}$	+4,200	+1,330	−6,195	−315
Change in T_m	+12°	+1°	−28°	−24°

Thermal transition temperature (T_m) in the absence of salt: 61.5°C.
Nandi, P.K. & Robinson, D.R. (1972) *J. Am. Chem. Soc.,* **94**, 1308.

tures of thermal unfolding were observed to increase. The action of various denaturants can thus be explained in terms of the transfer free energy changes of side chains and peptides (Δg_t).

4) For thermal unfolding of a protein, the change in free energy $\Delta G(T)$ of unfolding at a given temperature T can be evaluated using the equation

$$\Delta G(T) = \Delta H(T) - T\Delta S(T)$$
$$= \Delta H(T_m) + \Delta C_p(T - T_m) - T\Delta S(T_m) - T\Delta C_p\left(\frac{T}{T_m}\right). \quad (5.34)$$

Since $\Delta G = 0$ at $T = T_m$

$$\Delta S(T_m) = \frac{1}{T_m}\Delta H(T_m). \quad (5.35)$$

By substituting this equation into Eq. (5.34), the following equation is obtained.

$$\Delta G(T) = \Delta H(T_m)\left(1 - \frac{T}{T_m}\right) - \Delta C_p\{(T_m - T) + T \ln\left(\frac{T}{T_m}\right)\} \quad (5.36)$$

To obtain the change in free energy, $\Delta G(T)$, at temperature T, it is necessary to know the values of T_m, $\Delta H(T_m)$ and ΔC_p. The values of T_m and $\Delta H(T_m)$ can be obtained by measuring the thermal unfolding curve or by calorimetry. The value of ΔC_p can be obtained by calorimetry, or by the following method. Thermal unfolding curves of a protein are measured at various pH values. The T_m value varies depending on pH. Assuming a two-state transition, the changes in $\Delta H(T_m)$ at various T_m values on unfolding are determined using a van't Hoff plot. From the slope of the dependence of $\Delta H(T_m)$ on T_m, the change in heat capacity on unfolding can be estimated. The following method is also used to determine the heat capacity change on unfolding (Pace & Laurents, 1989). T_m and ΔH_m values are determined using a thermal unfolding curve and $\Delta G_D^{H_2O}$ is determined using an unfolding curve based on urea or guanidine hydrochloride. Then, ΔC_p can be calculated from T_m, ΔH_m and $\Delta G_D^{H_2O}$ using Eq. (5.36). The changes in heat capacity on unfolding of various proteins are shown in Table 5-10. Except for haem proteins such as cytochrome c and myoglobin, the average value of ΔC_p per residue is found to be 12 cal/deg/mol.

TABLE 5-10
Changes in Heat Capacity (ΔC_p) upon Unfolding of Various Proteins

Protein	ΔC_p (cal/deg/mol)	Protein	ΔC_p (cal/deg/mol)
Ribonuclease A	1,290[a]	Nuclease	2,180[a]
	2,200[c]	Carbonic anhydrase	3,930[a]
Hen lysozyme	1,590[a]	Ribonuclease T1	1,650[c]
T4 phage lysozyme	2,180[b]	C_L fragment	1,800[d]
β-Trypsin	3,070[a]	Cytochrome c	1,770[a]
α-Chymotrypsin	2,180[a]	Myoglobin	2,720[a]

[a]Privalov, P.L. & Gill, S.S. (1988) *Adv. Protein Chem., 39*, 191.
[b]Chen, B.-lu & Schellman, J.A. (1989) *Biochemistry, 28*, 685.
[c]Pace, C.N. & Laurents, D.V. (1989) *Biochemistry, 28*, 2520.
[d]Okajima, T. *et al.* (1990) *Biochemistry, 29*, 9168.

The value of ΔG at 25°C estimated using Eq. (5.36) agrees well with the value of $\Delta G_D^{H_2O}$ estimated using Eq. (5.28), (5.25) or (5.33).

5-5. METHODS OF INCREASING THE STABILITY OF PROTEINS

Site-specific modification by synthetic and recombinant DNA technology has recently been used to clarify the structure, stability and function of proteins. This method is useful for understanding the contribution of each amino acid residue to structure and stability. As described above, the difference in the free energy between the native and unfolded states is very small, because the stabilizing contributions of different types of non-covalent interaction counteract the destabilizing contribution of the conformational entropy of the polypeptide chain (Section 5-4). Therefore, if non-covalent interactions, conformational entropy or both are changed slightly by replacement of one or more residues by site-directed mutagenesis, protein stability would be expected to change greatly. The following methods have been used to increase the protein stability.

A) Modification of Conformational Entropy

For a protein of given size, the number of conformations of the unfolded state when disulphide bonds are present is smaller than when they are absent. Therefore, reduction of disulphide bonds increases the change in conformational entropy on unfolding and thus decreases the protein's stability. According to Flory (1956), the decrease in the conformational entropy of a randomly coiled polypeptide chain associated with formation of cross-links such as disulphide bonds is given by the following equation

$$\Delta S = 0.75 \ \nu R(\ln \frac{n}{\nu} + 3), \tag{5.37}$$

where R is the gas constant, ν is twice the number of bridges, and n is the number of residues in a loop formed by a bridge.

The following equation has also been proposed to estimate the decrease in entropy on formation of disulphide bonds (Pace *et al.*, 1988)

$$\Delta S = -\frac{3}{2}R \ln n - 2.1. \tag{5.38}$$

This equation was derived by calculating the probability of finding the ends of two chains in a volume element v_s. When two SH groups form a disulphide bond, they must approach within a distance of about 4.8 Å, and it is assumed that v_s can be approximated by a sphere 4.8 Å in diameter (57.9 Å^3).

The C_L or V_L domain can be isolated as a fragment by limited proteolysis of the immunoglobulin light chain. Each of the C_L and V_L fragments has only one intrachain disulphide bond buried in the interior of the molecule. Most proteins with disulphide bridges contain two or more, and proteins with only one disulphide bond are rare. Thus the C_L and V_L fragments are well suited to clarifying the role of the disulphide bond in protein stability. When the disulphide bond of a C_L fragment has been reduced in 4 M guanidine hydrochloride or 8 M urea, and the denaturing agent and reducing agent have been removed by dialysis, a C_L fragment with a reduced disulphide bond can be obtained. Various physico-chemical studies have shown that no great conformational change occurs on reduction of the disulphide bond. However, the reduced C_L fragment is less resistant than the intact C_L fragment to guanidine hydrochloride. From analysis of the unfolding curves by guanidine hydrochloride using Eq. (5.25), the changes in free energy of unfolding ($\Delta G_D^{H_2O}$) in the absence of denaturant are estimated to be 5.7 kcal/mol for the intact C_L fragment and 1.7 kcal/mol for the reduced C_L fragment. This indicates that reduction of the disulphide bond destabilizes the C_L fragment by 4 kcal/mol (Goto & Hamaguchi, 1979).

For the following cycle in water

$$
\begin{array}{ccc}
\text{Intact } C_L & \xrightarrow{\ (1)\ } & \text{Intact } C_L \\
(N) & & (D) \\
(3) \downarrow & & \downarrow (2) \\
\text{Reduced } C_L & \xrightarrow[\ (4)\ ]{} & \text{Reduced } C_L \\
(N) & & (D)
\end{array}
\tag{5.39}
$$

the relation $\varDelta G_1 + \varDelta G_2 = \varDelta G_3 + \varDelta G_4$ holds. Experimentally, $\varDelta G_1$ and $\varDelta G_4$ were determined to be 5.7 and 1.7 kcal/mol, respectively.

If we assume that no conformational change occurs on reduction of the intrachain disulphide bond and that there is no strain in the disulphide bond, $\varDelta S_3$ equals zero and $\varDelta H_3$ and $\varDelta H_2$ are the same. Therefore, the difference in the value of $\varDelta G$ between processes (1) and (4) may be written as

$$
\varDelta G_1 - \varDelta G_4 = T\varDelta S_2.
\tag{5.40}
$$

This indicates that the decreased stability of the reduced C_L fragment compared with the intact C_L fragment may be explained in terms of the increased entropy of the C_L fragment following reduction of the intrachain disulphide bond in the unfolded state.

There are 60 residues in the loop formed by the disulphide bond in the C_L fragment. Therefore, the entropy change ($\varDelta S_2$) is estimated to be 19 and 14 cal/deg/mol by use of Eqs. (5.37) and (5.38), respectively. This contributes to the free energy by 5.7 and 4.3 kcal/mol, respectively, these values being comparable with the value of ($\varDelta G_1 - \varDelta G_4$) (4 kcal/mol). This indicates that the lower stability of the reduced C_L fragment is indeed due mostly to the larger entropy of the reduced C_L fragment compared with that of the intact C_L fragment in the unfolded state. A comparison of the kinetics of the intact C_L fragment with those of the reduced C_L fragment showed that the presence of the disulphide bond does not affect the refolding rate, but slows down the unfolding rate (Goto & Hamaguchi, 1982a,b). This example indicates that one of the methods for increasing the protein stability is to decrease the conformational entropy by forming cross-links in the unfolded state.

Oxidation of hen egg-white lysozyme with iodine results in formation of an ester linkage between Glu 35 and Trp 108. The stability of lysozyme is increased greatly by this esterification, and the increase is also explicable in terms of a decrease in conformational entropy in the unfolded state (Johnson *et al.,* 1978). Since Glu 35 is one of the catalytic groups of hen egg-white lysozyme, the catalytic activity is abolished by the esterification.

Two other examples may be cited. T4 phage lysozyme has two free cysteine residues at positions 54 and 97. Computer graphics analysis of the lysozyme conformation indicated that the distance between C_α atoms of residues Cys 97 and Ile 3 is such that formation of a disulphide bond is possible. Therefore, a protein in which Ile 3 was replaced by Cys using site-directed mutagenesis and a disulphide bond formed by oxidation between Cys 97 and Cys 3, was prepared. It was shown that this mutant protein had the same activity as the wild-type lysozyme, but with increased thermal stability. However, since this mutant protein also contained the free Cys 54, an interchange reaction occurred between the Cys 97-Cys 3 disulphide bond and the SH group of Cys 54 at high temperatures. In a mutant protein in which Cys 54 was replaced by Thr or Val, no interchange reaction occurred and the thermal stability was found to be further increased (Perry & Wetzel, 1986).

Subtilisin BPN′ has no cysteine residues. Pantoliano *et al.* (1987) attempted to introduce two Cys residues by site-directed mutagenesis and to form a disulphide bond between them. Analysis of the structure had shown that Thr 22 and Ser 87 were located at positions suitable for formation of a disulphide bond. They prepared a mutant protein in which these residues were replaced by Cys, and a disulphide bond was formed between them. The midpoint temperature (T_m) of the thermal unfolding curve was higher by 3.1°C than that of the wild-type protein and by 5.8°C than that of a protein in which the disulphide bond was reduced. The mutant protein had the same activity as the wild-type protein. The rate of thermal inactivation of the mutant protein was about half that of the mutant protein. The increased stability achieved by introduction of the disulphide bond in the subtilisin BPN′ molecule was also explicable in terms of the decreased entropy in the unfolded state.

There are, however, certain problems regarding this method of stabilizing the protein molecule by the introduction of cross-links. The first concerns the strain in the disulphide bond that is introduced. Most disulphide bonds in wild-type proteins have strain energies of dihedral angles between 0.5 and 1.7 kcal/mol (Katz & Kossiakoff, 1986). The strain energy of the disulphide bond of each domain in the immunoglobulin molecule is estimated to be 1.18 kcal/mol. These strains act to destabilize the protein molecule. The strain energy of the disulphide bond introduced between Cys 22 and Cys 87 in subtilisin BPN′ was estimated to be 4.79 kcal/mol, which is larger than that of a disulphide bond introduced between Cys 24 and Cys 87 (2.46 kcal/mol). In fact, the mutant protein with the disulphide bond introduced between Cys 24 and Cys 87 had greater stability than the mutant protein with the Cys 22-Cys 87 disulphide bond. The strain energy of a disulphide bond introduced between Cys 39 and Cys 85 in dihydrofolate reductase was large (estimated to be 3.4–4.7 kcal/mol), and the stability of this mutant protein did not increase compared with the wild-type protein.

The second problem involves the local or global conformational change caused by introduction of a cross-link. As an extreme example, we will describe the conformation and stability of a derivative of the C_L fragment in which the disulphide bond is replaced by an S-Hg-S bond (Goto & Hamaguchi, 1986). The derivative was prepared by reacting the reduced C_L fragment with mercuric chloride. The mercury derivative was as compact as the intact C_L or reduced C_L fragment, and a tryptophyl residue was found to be buried near the S-Hg-S bond in the interior of the protein molecule. Judging from the circular dichroism spectrum, however, the β-structure characteristic of the immunoglobulin-fold was disturbed. The stability of the derivative to guanidine hydrochloride was lower than that of the intact C_L fragment and the change in free energy of unfolding ($\Delta G_D^{H_2O}$) in water was estimated to be 1.4 kcal/mol. This value is close to the value of $\Delta G_D^{H_2O}$ for the

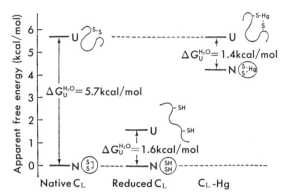

Fig. 5-8.   Free energy profiles for unfolding of intact C_L, reduced C_L and mercury derivative of C_L (pH 7.5, 25°C).

reduced C_L fragment (1.7 kcal/mol, see p. 143). Since we can assume that the conformational entropy of the mercury derivative is the same as that of the intact C_L fragment in the unfolded state, the decreased stability of the mercury derivative must be due to a change in conformation caused by replacement of the S–S bond by the S–Hg–S bond in the folded state. Therefore, the apparent similarity of the $\Delta G_D^{H_2O}$ values for the mercury derivative and the reduced C_L fragment cannot be attributed to the same cause. As illustrated in Fig. 5-8, the energy level of the reduced C_L fragment is the same as that of the intact C_L fragment in the folded state, because the conformations of both proteins are the same as described above. On the other hand, the energy level of the mercury derivative is the same as that of the intact C_L fragment in the unfolded state, because both proteins have a cross-link between the same positions.

Finally, attention should be given to chemical modification of disulphide bonds at alkaline pH values or at high temperatures. A disulphide bond is subjected to β-elimination by alkali, to yield one thiol and one dehydroalanine, and the latter reacts with the side chain of a Lys residue to form a cross-link. This results in irreversible thermal unfolding of proteins.

$$
\begin{array}{ccc}
\begin{array}{c}
-\text{CH}- \\
| \\
\text{CH}_2 \\
| \\
\text{S} \\
| \\
\text{S} \\
| \\
\text{CH}_2 \\
| \\
-\text{CH}-
\end{array}
&
\longrightarrow
&
\begin{array}{cl}
-\text{C}- & \\
\| & \\
\text{CH}_2 & \text{(dehydroalanine)} \\
+ & \\
\text{SH} & \\
| & \\
\text{S} & \\
| & \\
\text{CH}_2 & \\
| & \\
-\text{CH}- &
\end{array}
\end{array}
$$

$$
\overset{|}{\underset{|}{\text{C}}}=\text{CH}_2 \; + \; \text{NH}_2-(\text{CH}_2)_4-\overset{|}{\underset{|}{\text{CH}}} \longrightarrow \; \text{H}\overset{|}{\text{C}}-\text{CH}_2-\text{NH}-(\text{CH}_2)_4-\overset{|}{\underset{|}{\text{CH}}}
$$
$$
\qquad\qquad\qquad\qquad \text{Lys}
$$

Replacement of one residue with a more bulky residue also decreases the conformational entropy in the unfolded state (Matthews *et al.*, 1987). For instance, a mutant protein in which Gly 77 is replaced by Ala and one in which Ala 87 is replaced by Pro have the same crystallographic structure as the wild-type T4 phage lysozyme, but show slightly increased thermal stability. This was explained in terms of a decrease in conformational entropy in the unfolded state caused by replacement of Gly (or Ala) with a more bulky residue, Ala (or Pro). Hecht *et al.* (1986) and Imanaka *et al.* (1986) also observed an increase in thermal stability by replacement of Gly with Ala in the λ repressor and the neutral protease of *Bacillus stearothermophilus*. However, they (probably erroneously) interpreted this finding in terms of the replacement of Gly, which has no helix-forming propensity, with Ala, which has a strong helix-forming propensity.

B) *Modification of Non-covalent Interactions*

The method of stabilizing the protein molecule by decreasing the conformational entropy in the unfolded state has been described above. On the other hand, there have been many attempts to increase the stability of the protein molecule by increasing the interactions of hydrogen bonds and hydrophobic bonds in the folded state through replacement of one or more residues by site-directed mutagenesis. As described below, however, replacement of one residue occasionally causes an unexpectedly large change in the protein conformation, and the change in stability cannot be explained only by assuming that the orientations of the backbone and side chains of a mutant protein are the same as those of the wild-type protein determined by X-ray crystallography. In this regard, studies by Matthews' group in Oregon have revealed very interesting findings. They studied changes in the stability of various mutants of T4 phage lysozyme, while at the same time investigating the crystallographic structure of each mutant protein.

Alber *et al.* (1987b) studied the crystallographic structures and the changes in thermal stability of mutant proteins of T4 phage lysozyme in which Thr 157 was replaced by 13 different amino acid residues by site-directed mutagenesis. Changes in the thermal stability of these 13 different mutant proteins are shown in Table 5-11. Thr 157 is located in an irregular loop on the surface of the protein molecule. Replacement of this residue with each of 12 different residues gave a mutant with stability between that of the wild-type protein and a temperature-sensitive (*ts*) mutant. Changes in stability caused by these replacements are relatively small, but it is

TABLE 5-11
Changes in Thermal Transition Temperature (ΔT_m) and Instability ($\Delta \Delta G_D^{H_2O}$) of T4 Phage Lysozyme Caused by Replacement of Thr 157 by Various Amino Acid Residues

Amino acid residue at position 157	T_m (°C)	$\Delta \Delta G_D^{H_2O}$ (kcal/mol)
Thr (wild type)	—	—
Asp	−1.7	0.45
Ser	−2.5	0.66
Asp	−4.2	1.1
Gly	−4.2	1.1
Cys	−4.9	1.3
Leu	−5.0	1.3
Arg	−5.1	1.3
Ala	−5.4	1.4
Glu	−5.8	1.5
Val	−6.0	1.6
His	−7.9	2.1
Phe	−9.2	2.4
Ile (*ts* mutant)	−11.0	2.9

Alber, T. *et al.* (1987) *Nature, 330,* 41.

interesting to note that Thr 157 can tolerate this replacement with various residues. The OH group of Thr 157 forms a hydrogen bond between the peptide NH of Asp 159 (Fig. 5-9). Replacement with residues (Asn, Ser, Asp and Gly) that can retain this hydrogen bond did not cause a great decrease in stability (Fig. 5-10), but it was found that replacement of Thr 157 with some other residues gave a local but unexpected conformational change. For instance, a water molecule is newly bound in a mutant protein bearing Gly 157 (Fig. 5-10 and Table 5-12). In a mutant protein bearing Asp 157, the Asp residue has two orientations; one is the same as that of Thr 157 in the wild-type protein, and in the other the side chain of Asp 157 rotates and is oriented in the opposite direction. In some mutants, a water molecule was found to be bound to Thr 155 and Asp 159. An ion-pair is newly formed in a mutant with Arg 157. In a mutant bearing Ser 157, the movement of the OH group of the Ser residue is greater than that of the OH group of Thr 157 in the wild-type protein. These changes were not predicted from X-ray crystallographic analysis of the wild-type protein alone.

Matsumura *et al.* (1988a) studied the X-ray crystallographic structures and stability of mutants of T4 phage lysozyme in which Ile 3 was replaced by various amino acid residues. Ile 3 is surrounded by the side chains of Met 6, Leu 7 and Ile 100 and the main chain of Cys 97, which are buried in the interior of the molecule, and has a relative accessibility of about 20%. A mutant protein in which Ile 3 is replaced by Val showed no conformational change except for absence of the δ-CH_3 of the Val side chain. However, a substantial conformational change was observed for a mutant bearing Tyr 3; the side chain of Tyr 3 is not accommodated in the interior of the molecule and is frequently exposed to solvent. Thermal unfolding of mutant lysozymes in which Ile 3 was replaced by various amino acid residues was studied, and the changes in the stability ($\Delta\Delta G_D^{H_2O}$) at T_m of the wild-type lysozyme were determined using Eq. (5.24). In Fig. 5-11, the values of $\Delta\Delta G_D^{H_2O}$ thus obtained are plotted against the hydrophobic scales of the side chains (Δg_t) (Table 1-4). As can be seen, a linear relation was obtained except for mutants in which Ile 3 was replaced by Phe,

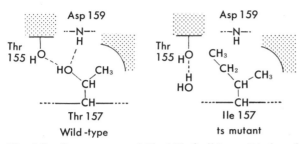

Fig. 5-9. Structure around Thr 157 of wild-type T4 phage lysozyme and structure around Ile 157 of the temperature-sensitive (*ts*) mutant of the lysozyme. (Alber, T. *et al.* (1987) *Nature, 330,* 41)

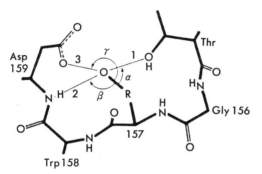

Fig. 5-10. Hydrogen-bonded network formed by various replaced amino acid residues at position 157 of T4 phage lysozyme. (Alber, T. *et al.* (1987) *Nature, 330,* 41)

TABLE 5-12
Hydrogen-bonded Network of Various Amino Acid Residues at Position 157 of T4 Phage Lysozyme (see Fig. 5-10)

Amino acid residue	Atom	Atomic distance (Å)			Angle (°)			B factor (Å²)
		1	2	3	α	β	γ	
Thr	O_γ	2.8	3.2	3.2	119	103	106	20
Ser	O_γ	2.8	3.4	3.2	119	107	114	30
Asn	O_δ	3.1	3.1	3.2	97	130	101	30
Asp(1)	O_δ	2.9	3.1	3.1	100	125	94	—
Gly	H_2O	2.9	3.1	2.8	—	—	—	29

Alber, T. *et al.* (1987) *Nature, 330,* 41.

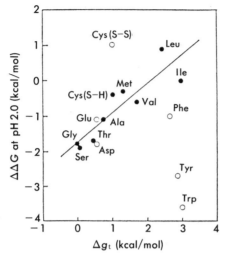

Fig. 5-11. Changes in stability ($\Delta\Delta G_D^{H_2O}$) caused by replacement of Ile 3 of T4 phage lysozyme by various amino acid residues plotted against the hydrophobicity (Δg_t) of amino acid side chains. (Matsumura, M. *et al.* (1988) *Nature, 334*, 406)

Tyr or Trp. This indicates that the protein stability increases as the hydrophobicity of the replaced residue increases. The deviations observed for the mutant proteins containing Phe, Tyr and Trp may be due to lack of accommodation of these bulky side chains in the interior of the molecule.

Yutani *et al.* (1987) studied the unfolding by guanidine hydrochloride of mutant proteins of the α-subunit of tryptophan synthase in which Glu 49 was replaced by 20 different amino acid residues. The changes in free energy ($\Delta G_D^{H_2O}$) in water estimated using Eq. (5.25) are plotted against the hydrophobicity of the side chains of residues replaced (Δg_t) in Fig. 5-12. As observed for mutants of T4 phage lysozyme, the stability increases as the hydrophobicity of the replaced residues (except Phe, Tyr and Trp) increases.

Matsumura *et al.* (1988b) studied the heat inactivation of mutant proteins of kanamycin nucleotidyltransferase in which Asp 80 was replaced by various residues, and found that the half-life

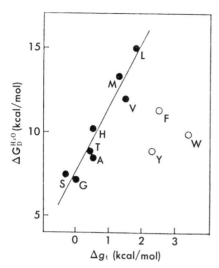

Fig. 5-12. Stability ($\Delta G_D^{H_2O}$) of the mutants of the α-subunit of tryptophan synthase in which Gln 49 is replaced by various amino acid residues plotted against the hydrophobicity (Δg_t) of amino acid side chains. (Yutani, K. *et al.* (1987) *Proc. Natl. Acad. Sci. U.S.A.*, *84*, 4441)

values for inactivation correlated well with the hydrophobicity of the replaced residue.

Pro 86 in T4 phage lysozyme is located in an α-helix and causes a kink in the helix. Replacement of this residue by ten different residues, Arg, Ala, Asp, Gly, His, Ile, Leu, Ser and Thr, causes a significant conformational change in the peptide from residues 81 to 83, and a continuous helix is formed between residues 81 and 91. In addition, a small conformational change over a distance of 20 Å occurs. Nevertheless, no appreciable change in thermal stability was observed as a result of these replacements, and the change in free energy on unfolding was only decreased by less than 0.5 kcal/mol when compared at T_m of the wild-type lysozyme (Alber *et al.*, 1988).

Replacements of single amino acid residues in 25 different temperature-sensitive mutants of T4 phage lysozyme were found to occur at sites with small crystallographic temperature factors (B-factor) (see Section 6-2) and with small accessible surface areas

(Alber *et al.*, 1987a) (Fig. 5-13). This indicates that an amino acid residue which is fixed in the interior of the protein molecule contributes greatly to the stability by specific interactions with the surrounding atoms, and that an exposed residue with high mobility does not contribute significantly to the protein stability.

Replacement of Gly 156 in T4 phage lysozyme by Asp decreased T_m by 6.1°C (Gray & Matthews, 1987). Replacement of Asn 57 in iso-1-cytochrome c by Ile increased the thermal transition temperature by 17°C, which corresponds to an increase in $\Delta G_D^{H_2O}$ of 4.2 kcal/mol (Das *et al.*, 1989). This is the largest enhancement of protein stability attained so far by replacement of a single amino acid residue.

The T_m value of the N-terminal fragment of the wild-type λ repressor is 54°C. Replacement by Gly of Leu 57, which is located in the hydrophobic core of the fragment, decreased the T_m value by 50°C (Sauer *et al.*, 1990). Replacement of Gly 48 by Ala or Gly 46 by Ala increased the T_m value by 4.7°C and 3.1°C, respectively. It is not clear whether this increase in the stability is due to the high helix-forming propensity of Ala or to decreased conformational entropy of the unfolded mutant protein or formation of the structure in some other way.

C) Interactions between Helix Dipole and Charge

As described in Section 2-4, interactions between an α-helix dipole and a negative charge at the N-terminus or a positive charge at the C-terminus stabilize the α-helix. Nicholson *et al.* (1988) attempted to enhance the stability of T4 phage lysozyme by introducing a negative charge at the N-terminus of an α-helix. There are eleven α-helices in the lysozyme molecule; seven of these have a negative charge at their C-terminus in the wild-type protein. Nicholson and his colleagues prepared a mutant protein (S38D) in which a negative charge was introduced into the α-helix from residue 39 to 50 by replacement of Ser 38 by Asp, and a mutant protein (N144D) in which a negative charge was introduced by replacement of Asn 144 by Asp in the α-helix from residue 143 to

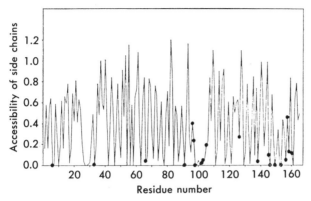

Fig. 5-13. The accessibility of the side chains of T4 phage lysozyme plotted against the residue number. Black dots indicate the positions at which amino acid residues are replaced in each temperature-sensitive mutant. (Alber, T. *et al.* (1987) *Biochemistry,* 26, 3754)

155. The thermal transition temperatures of S38D and N144D were both increased by 2°C compared with those of the wild-type protein. The thermal transition temperature of a mutant protein in which Ser 38 and Asn 144 were both replaced by Asp was found to increase by 4°C, which corresponds to an increase in the free energy of stabilization of 1.6 kcal/mol. X-ray crystallographic studies have shown that Ser 38 in the wild-type protein forms hydrogen bonds with the amide nitrogens of residues 40 and 41, and that similar hydrogen bonds are also formed in the S38D mutant, although the geometry of these hydrogen bonds is poorer than in the wild-type protein. Neither Asn 144 in the wild-type lysozyme nor Asp 144 in the N144D mutant has direct hydrogen bonding with the end of the helix. These structural data suggest that the stabilization of the S38D and N144D mutants is indeed due to electrostatic interaction between the helix dipole and the negative charge at the C-terminal end, and not due to hydrogen bonding between the replaced residue and the end of the helix. Similar studies were also carried out on barnase (*Bacillus amyloli-quefaciens* ribonuclease) by Săli *et al.* (1988).

D) Chemical Modification

We have described the changes in stability that occur upon replacement of one or more residues. It has been shown that protein stability may be greatly changed, even if no conformational change occurs through replacement of residues. In contrast to substitution of a whole residue, however, chemical modification can change only a small part of a residue. Studies on changes in stability by chemical modification are thus interesting to clarify further structure-stability relationship within the protein molecule.

As an example, the effects of chemical modification of tryptophan residues on protein stability may be cited (Okajima *et al.*, 1990). No other amino acid closely resembles tryptophan. It is thus impossible to make a small change in a tryptophan side chain by site-directed mutagenesis, and this can only be achieved by chemical modification. Ozone oxidation modifies the tryptophan residue to N'-formylkynurenine (NFK) (Kuroda et al., 1975), which is converted to kynurenine (Kyn) by freezing in acid (Yamasaki *et al.*, 1979).

Trp   NFK   Kyn

X-ray crystallographic analysis of N-acetylkynurenine crystals has shown that the carbonyl oxygen is hydrogen-bonded to the aromatic amino group at the ortho-position, that this hydrogen-bonded ring is coplanar within 3° with the benzene ring in the Kyn, and that this structure is similar to the structure of tryptophan (Kennard *et al.*, 1979).

The C_L fragment of the type-λ immunoglobulin light chain has two tryptophan residues at positions 150 and 187. X-ray crystallographic studies have shown that Trp 150 is buried completely in the interior and Trp 187 is located near the surface of the molecule. Oxidation of Trp 187 by ozone to NFK or Kyn produced a large decrease in the stability to guanidine hydrochloride and thermal denaturation, although the conformation was not changed. The thermal transition temperatures and the values of $\Delta G_D^{H_2O}$ were found to be 61.0°C and 4 kcal/mol for the intact C_L fragment, 46.1°C and 2 kcal/mol for the C_L fragment in which Trp 187 was modified to NFK, and 48.8°C and 2 kcal/mol for the C_L fragment in which Trp 187 was modified to Kyn.

Okajima *et al.* (1990) also studied the changes in stability of ribonuclease T1 and hen egg-white lysozyme caused by ozone oxidation of tryptophan residues. Ribonuclease T1 has only one tryptophan residue at position 59, which is buried in the interior of the molecule. Modification of Trp 59 resulted in a remarkable decrease in stability ($T_m = 48.7$°C and $\Delta G_D^{H_2O} = 5$ kcal/mol for intact ribonuclease T1, $T_m = 25.3$°C and $\Delta G_D^{H_2O} = 0.1$ kcal/mol for ribonuclease T1 in which Trp 59 was modified to NFK, and $T_m = 33.9$°C and $\Delta G_D^{H_2O} = 2$ kcal/mol for ribonuclease T1 in which Trp 59 was modified to Kyn). Hen egg-white lysozyme has six tryptophan residues, and the degree of exposure of Trp 62 is the greatest. No significant change in stability occurred on modification of Trp 62 to NFK or Kyn (in the presence of 1.5 M guanidine hydrochloride at pH 7.5, $T_m = 57.9$°C for intact lysozyme, $T_m = 54.8$°C for lysozyme in which Trp 62 was modified to NFK, and $T_m = 57.6$°C for lysozyme in which Trp 62 was modified to Kyn).

As described above, the extent of the decrease in the stability of the proteins on modification of Trp increases in the order Kyn 62-lysozyme < Kyn 187-C_L fragment < Kyn 59-ribonuclease T1. The degree of exposure of the tryptophan residues estimated from the crystallographic thermal factors increases in the order Trp 59 (ribonuclease T1) < Trp 187 (C_L) < Trp 62 (lysozyme). Thus, the decrease in stability is closely related to the degree of exposure of the modified tryptophan residues; the lower the mobility or solvent accessibility of the tryptophan residue, the greater the extent of the decrease in stability upon modification. The results described here show how delicately the interactions of a given residue with surrounding atoms are balanced in the protein molecule. It must be kept in mind that a small modification of a side chain will generally cause a great change in protein stability.

E) Binding of Metal Ions

Binding of a metal ion to a specific site enhances protein stability. For instance, the presence of 0.1 M NaCl, MgCl$_2$ and Na$_2$HPO$_4$ increases the free energy of stability of ribonuclease T1 by 0.8, 1.8 and 3.3 kcal/mol, respectively (Pace & Grimsley, 1988). These increases in stability result from the binding of a Na$^+$ or Mg^{2+} ion to a cation binding site and a HPO$_4^{2-}$ ion to an anion binding site. The binding constants for Na$^+$, Mg^{2+} and HPO$_4^{2-}$ are 62, 155 and 282 M^{-1}, respectively. These results suggest that design of a specific binding site for a cation or anion by replacement of amino acids on the surface of a protein molecule is one further method for enhancing protein stability.

For a single binding site, the increase in the stability ($\Delta\Delta G$) due to ion binding is given by

$$\Delta\Delta G = \Delta G(\text{in the presence of ion}) - \Delta G(\text{in the absence of ion})$$
$$= RT \ \ln(1 + Ka_{\pm}) \tag{5.40}$$

where K is the binding constant and $a_{\pm}$ is the mean ionic activity of the ion (Schellman, 1975). Thus the increase of stability depends on the binding constant and the concentration of the ion.

α-Lactalbumin has a Ca^{2+}-binding site which consists of the carbonyl oxygens of Lys 79 and Asp 84, the carboxylate groups of Asp 82, Asp 87 and Asp 88, and two water oxygens (Acharya et al., 1989). Binding of a single Ca^{2+} ion to the high-affinity site of α-lactalbumin was found to increase the unfolding transition temperature (Mitani et al., 1986). Human lysozyme has no binding site for Ca^{2+} ion. Kuroki et al. (1989) prepared a mutant of human lysozyme in which both Gln 86 and Ala 92 were replaced by Asp using site-directed mutagenesis. It was found that this mutant enzyme bound one mole of Ca^{2+} with a binding constant of 5.0 ×

10^6 M^{-1}, and the Ca^{2+}-binding mutant enzyme was much more stable than wild-type lysozyme.

Subtilisin BPN' has a high-affinity site and a low-affinity site for Ca^{2+} ion. Pantoliano *et al.* (1988) attempted to increase the binding constant for the low-affinity site, and thus enhance the stability of the subtilisin molecule by replacement of Pro 172, Gly 131 or both by Asp. These residues are located near the low-affinity site. Replacement by Asp of both Pro 172 and Gly 131 was found to increase the binding constant of Ca^{2+} about 6-fold over that of the wild-type protein and increase the stability.

Despite these results, the binding of ions or ligands is a somewhat tedious method for enhancing the thermal stability of proteins, and the binding constant may also change according to temperature.

F) Solvents that Stabilize the Protein Molecule

The stability of the protein molecule can be increased not only by modifying the molecule but also by changing its environment. Addition of $(NH_4)_2SO_4$, Na_2SO_4 (see Section 5-4), glycerol or sucrose (Timasheff & Arakawa, 1989) to the solvent is one method of stabilizing the protein molecule.

We have described various factors that enhance the stability of a protein. However, a change in stability caused by replacement of one or more amino acid residues cannot be attributed unequivocally to only one among the various phenomena described above: conformational change, modifications of hydrogen bonds and hydrophobic interactions, change in helix-forming propensity, change in helix dipole-charge interactions and conformational entropy in the unfolded state. Neither can modification of enzyme activity caused by amino acid replacement be attributed to only one cause. There have now been many studies that involve replacement of residues in the active site by other residues in order to test whether the enzyme mechanism proposed for the wild-type enzyme is correct. For instance, the hydroxyl group of Tyr 248 of carboxypeptidase A was postulated to be the donor of protons to the NH group of the substrate in the cleavage step. To test this postulate, a mutant enzyme, in which Tyr 248 was replaced by Phe, was prepared (Gardell *et al.*, 1985, 1987). It was found that the OH group of Tyr 248 participated in the substrate binding but not in catalysis. However, it is possible that in the mutant enzyme a water molecule may occupy the position of the OH group of Tyr 248 and participate in catalysis. Thus it is important to examine the X-ray crystallographic structure of the mutant enzyme if possible.

6
Protein Dynamics

As described in Chapter 3, the interior of the protein molecule has a compact structure. However, there is much evidence to suggest that proteins, in fact, have fluctuating dynamic structures. It is often observed that there are regions where electron densities are not well defined in protein crystals, indicative of large-scale movement of a polypeptide segment. The following phenomena also suggest fluctuations in the interior of the protein molecule. (a) Oxygen molecules with no charge penetrate rapidly into the interior of the protein molecule and quench the fluorescence of tryptophan residues. (b) Various groups in the interior are modified by reagents added to the medium. (c) Amino acid side chains rotate rapidly in the interior of the molecule (flip-flop motion detected by NMR). (d) The hydrogens of NH, SH and OH groups of peptides and side chains buried in the interior of the molecule are replaced by deuterium when proteins are dissolved in heavy water.

6-1. LARGE-SCALE MOVEMENT OF POLYPEPTIDE SEGMENTS

An interesting example of large-scale intramolecular movements can be found in trypsinogen and trypsin (Huber & Bode,
1978). Trypsinogen is the precursor of trypsin. Specific cleavage of
the peptide bond between residues 15 and 16 of trypsinogen yields
active trypsin with Ile 16 at the N-terminus. The structure of trypsin
is shown in Fig. 6-1.

X-ray crystallographic analysis has shown that about 85% of
the polypeptide chain of trypsinogen and trypsin have the same
conformation, whereas the remainder differs greatly. The part that

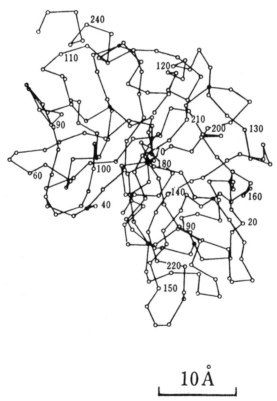

10 Å

Fig. 6-1. The structure of trypsin. (Depicted by Y. Mitsui based on the coordinates of Huber *et al.* (1974) *J. Mol. Biol., 89.,* 73)

exhibits a different conformation consists of four segments, N-terminus–Gly 19, Gly 142–Pro 152, GlyA 184–Gly 193, and Gly 216–Asn 223, and is called the activation domain. This part has a definite conformation for trypsin but not for trypsinogen. All these four segments have Gly residues at their N-termini: Gly with no side chain seems to be suited for chain flexibility. No aromatic residues are present in the activation domain, nor in the flexible switch or hinge regions of the antibody molecule. Aromatic residues seem to confer "stiffness" to the polypeptide chain.

More than 20 hydrogen bonds, and in particular the ion-pair between the COO$^-$ group of Asp 49 and the NH$_3^+$ group of Ile 16 are important for stabilizing the activation domain of trypsin (Fig. 6-2); the activation domain of trypsinogen is in part unstable due to the lack of this ion-pair. The disulphide bond between Cys 191 and Cys 220 in the activation domain is easily reduced for trypsinogen but not for trypsin. The segments of the activation domain of trypsin form the substrate-binding site, and Asp 189 gives trypsin its specificity through interactions with Lys or Arg of the substrate molecule.

On binding of bovine pancreatic trypsin inhibitor to trypsino-

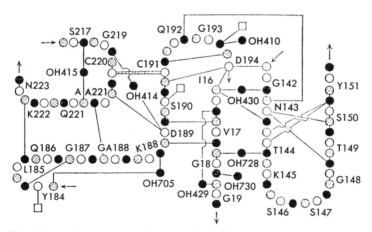

Fig. 6-2. The activation domain of trypsin. ⊘, N; ○, C; ●, O; —, hydrogen bonds between residues of the activation domain; —□, hydrogen bonds to residues outside the domain; —●, hydrogen bonds mediated by immobilized water. (Huber, R. & Bode, W. (1978) *Acc. Chem. Res., 11,* 114)

gen, however, the conformation of the activation domain becomes ordered. Utilizing the binding energy of trypsin inhibitor, trypsinogen assumes a trypsin-like conformation. However, the segment from the N-terminus to Gly 18 in the activation domain of trypsinogen remains mobile. Several hydrogen bonds and ion-pairs are formed around Ile 16 in the pocket of trypsin, and when trypsin inhibitor binds to trypsinogen, a pocket is formed around Ile 16, but is empty. When a dipeptide, Ile-Val, is added to this complex, the dipeptide binds firmly to this pocket. The structure of the ternary complex of trypsinogen, trypsin inhibitor and dipeptide is indistinguishable from the binary complex of trypsin and trypsin inhibitor.

X-ray crystallographic analysis of myeloma proteins has shown that the electron density of the Fc portion is smeared out, although that of the Fab portion is well defined. However, the Fc portion of a myeloma protein which lacks the hinge region between C_H1 and C_H2 gives a well defined electron density map (Silverton *et al.*, 1977). The presence of the hinge region thus seems to be responsible for the disorder of the Fc portion.

T4 phage lysozyme consists of two domains. Recent X-ray studies on a mutant form of T4 lysozyme in which Met 6 is replaced by Ile have shown that the mutant crystallizes with five different conformations and that the hinge-bending angle varies by up to 32° (Faber & Matthews, 1990). This example also typifies the large-scale structural variability of the protein molecule.

6-2. FLUCTUATIONS IN THE PROTEIN MOLECULE

The mean square of the energy fluctuations, $\overline{\partial E^2}$, of a system is given by the following equation (Cooper, 1976)

$$\overline{\partial E^2} = kT^2 m C_v, \tag{6.1}$$

where k is the Boltzmann constant, T is the absolute temperature, m is the mass of the system and C_v is the specific heat capacity at constant volume.

For a single molecule with a molecular weight of 25,000 and specific heat capacity of 0.32 cal/deg/g, the root mean square of the energy fluctuations amounts to 6.4×10^{-20} cal/molecule, which corresponds to 36 kcal/mol on a molar basis. This value is larger than the free energy of stabilization of the protein molecule.

Similarly, the mean square of the volume fluctuations, $\overline{\partial V^2}$, is given by the equation

$$\overline{\partial V^2} = kTV\beta_\gamma, \qquad (6.2)$$

where β_γ is the compressibility at constant pressure.

If the compressibility of a protein is assumed to be 5×10^{-6} atm^{-1}, the root mean square of the volume fluctuations amounts to about 50 Å^3/molecule. This corresponds to the volume of two water molecules. Although the energy and volume fluctuations are estimated to be considerable, the correlation times (in the order of nanoseconds) are not long enough to induce unfolding of the native protein molecule.

A) Temperature Factors of Protein Crystals

The temperature factor B_i of a particular atom (i) under consideration is expressed by $B_i = 8\pi^2 \bar{u}_i^2/3 = 8\pi^2 u_x^2$, where $\bar{u}_i$ is the mean displacement of the atom along the normal to the reflecting planes and u_x is its x-component. The B factor gives a measure of localization of the atom in the crystal; the greater its value, the less localized the atom. The temperature factors in protein crystals are of the order of 12 to 20 Å^2, which corresponds to a mean displacement of between 0.15 and 0.5 Å.

The larger the relative accessibility of a side chain, the larger the value of the B factor. The B values of main chain atoms are small in the interior of the protein molecule and become greater as the main chain approaches the protein surface.

It is not easy to interpret B values in terms of flexibility, because it is necessary to distinguish between lattice disorder and internal flexibility. However, by performing crystallographic analysis at various temperatures between 220 and 300 K, Frauenfelder *et*

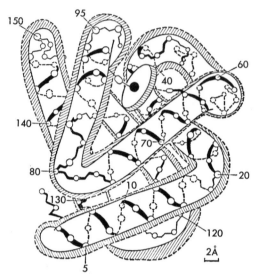

Fig. 6-3. The main chain structure of myoglobin. Solid lines indicate the static structure and the shaded area indicates the region reached by fluctuations with a 99% probability. Circles represent the α-carbons. (Frauenfelder, H. *et al.* (1979) *Nature, 280*, 558)

al. (1979) determined the conformational fluctuations of the main chain of myoglobin. In Fig. 6-3, the solid lines represent the static structure and the shaded area represents the region reached by fluctuations with a 99% probability at 250 K. Average displacements were found to be much larger for charged and neutral polar side chains than for non-polar side chains. In the haem environment, the displacements of side chains around distal His 64 are small and those around proximal His 93 are large. It was also shown that the fluctuations of the F-helix are large.

Specific protein-protein interactions seem to occur at sites with high fluctuations. Fontana *et al.* (1986) studied the cleavage sites of thermolysin by autolysis and also by limited digestion with subtilisin, which has broad substrate specificity, and found that there is a strong correlation between the sites of limited proteolysis and segmental mobility of the protein substrate. As shown in Fig. 6-4, in which the crystallographic B-factors are plotted against the polypeptide sequence, the sites of cleavage coincide with the

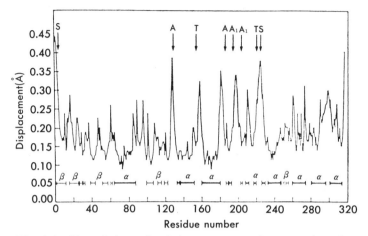

Fig. 6-4. Plot of the main chain temperature factors against the sequence of thermolysin. Bars at the bottom indicate segments of secondary structure (α, α-helix; β, β-structure). Arrows indicate sites of limited proteolysis or autolysis of thermolysin under different conditions. S, site of cleavage by subtilisin; A, cleavage by autolysis in the presence of 1.5 mM $CaCl_2$ and 1 mM EDTA; A_1, cleavage by autolysis in the presence of 1.5 mM $CaCl_2$ and 10 mM EDTA; T, cleavage by thermal autolysis. (Fontana, A. *et al.* (1986) *Biochemistry, 25*, 1847)

maximum mean-square displacements. It has been suggested that antigenic sites on proteins are regions of higher than average mobility (Tainer *et al.*, 1985; Davies *et al.*, 1988), but the X-ray structures of antibody-antigen complexes solved so far do not support this—they show "lock and key" interactions. It remains to be seen whether mobility is important here (Sutton, 1989; Davies & Padlan, 1990).

B) Flip-flop Motions

NMR studies have shown 180° rotations about the C_β-C_γ bond of Tyr and Phe rings, known as flip-flop motions. The structure of Tyr or Phe is symmetrical with respect to a two-fold axis through the C_β-C_γ bond. Therefore, if the side chain ring of the tyrosine residue is fixed rigidly in the interior of the protein molecule and the two hydrogen atoms at the ortho and meta positions are located in different environments, then these two hydrogens should show different resonance absorptions. However, when the ring of the Tyr residue rotates rapidly about the $C_\beta-C_\gamma$ bond, the two hydrogens will not be resolved in the NMR spectrum and their signals will be averaged. The extent of flip-flop motion depends on the environment in which the aromatic rings are located, and this motion is observed even when they are buried in the interior of the protein molecule. For instance, bovine pancreatic trypsin inhibitor has four Tyr residues at positions 10, 21, 23 and 35, with relative accessibilities of 0.32, 0.23, 0.04 and 0.06, respectively. Corresponding to the relative accessibility, the rings of Tyr 10 and Tyr 21 rotate very rapidly, the ring of Tyr 23 rotates about 5 times/sec at 4°C, and the ring of Tyr 35 rotates once/sec at 4°C, and about 6 times/sec at 27° C (Wüthrich & Wagner, 1978). The activation enthalpy for the flip-flop motions amounts to 20–25 kcal/mol. For the flip-flop motion to be possible in the interior of the protein molecule, a conformational change must occur, at least locally.

C) Reactivities of Groups in the Interior of the Protein Molecule

It has been observed that a group buried in the interior of the protein molecule can be modified by reagents added to the medium in the absence of denaturing agent. Unless the group is exposed by fluctuations and is accessible to the reagent in the solvent, the group will not be modified.

The C_L fragment of the immunoglobulin light chain contains only one disulphide bond. Although the accessible surface area of this disulphide bond is zero, the bond is reduced with dithiothreitol (DTT) even in the absence of denaturant (Kikuchi *et al.*, 1986). In order to be reduced with DTT, the disulphide bond of the C_L fragment must become accessible, and reduction by DTT may be explained on the basis of the following mechanism.

$$N \underset{k_2}{\overset{k_1}{\rightleftharpoons}} X \xrightarrow{k_3[\text{DTT}]} R, \qquad (6.3)$$

In this mechanism, N, the native C_L fragment containing a disulphide bond buried in the interior of the molecule and thus not accessible to solvent, is in equilibrium with X. X is any conformation of the C_L fragment molecule in which the disulphide bond is accessible to DTT. R is the C_L fragment in which the disulphide bond is reduced. k_1, k_2 and k_3 are the rate constants for the respective processes. The rate of reduction follows first-order kinetics. The reduction reaction of the disulphide bond with DTT was found to satisfy the condition, $k_1 \ll k_2 + k_3[\text{DTT}]$ and the dependence of the apparent first-order rate constant on DTT concentration was expressed by the equation

$$k_{\text{app}} = k_1 k_3 [\text{DTT}]/(k_2 + k_3 [\text{DTT}]). \tag{6.4}$$

The value of k_3 can be determined using the completely unfolded C_L fragment. The values of k_1 and k_2 thus determined at various concentrations of guanidine hydrochloride are in good agreement with the rates of unfolding and refolding. Furthermore, the values of k_1 and k_2 determined in the absence of guanidine hydrochloride are also close to the respective values, extrapolated to zero concentration of guanidine hydrochloride, of the unfolding and refolding rate constants determined at various concentrations of the denaturant. The values of k_1 and k_2 in the absence of denaturant are determined to be 7×10^{-5} sec^{-1} and 2.5 sec^{-1}, respectively.

The reduced C_L fragment has a conformation very similar to the intact C_L fragment, and the two SH groups are buried in the interior of the reduced protein (see p. 189). In 4 M guanidine hydrochloride, the SH groups react with 5,5'-dithiobis-(2-nitrobenzoic acid) (DTNB) instantaneously. In the absence of denaturant, the reaction of the SH groups with DTNB is rate-limited by the unfolding of the reduced C_L fragment. The unfolding rate was determined to be 1×10^{-3} sec^{-1}, which is considerably larger than the unfolding rate of the intact C_L fragment (7×10^{-5} sec^{-1}) (Goto & Hamaguchi, 1979).

The 33-kDa protein of the spinach oxygen-evolving complex has only one intrachain disulphide bond. Tanaka *et al.* (1989) studied the kinetics of reduction of the disulphide bond using DTT and found the bond reduced even in the absence of denaturant. The rate of reduction by DTT of the disulphide bond was 1,000 times slower than the reduction rate of the completely exposed bond. This suggests that the disulphide bond of the 33-kDa protein is located somewhere in the interior of the molecule and not exposed, and that the reduction is rate-limited by exposure of the disulphide bond as a result of fluctuations in the structure. The values of k_1 and k_2 for the reaction $N \rightleftharpoons X$ (Eq. (6.3)), and the unfolding and refolding rate constants for the reaction $N \rightleftharpoons U$ were determined at various concentrations of guanidine hydrochloride. It was found that the values of k_1 and k_2 obtained by the reduction kinetics are much larger than the respective values determined by global unfolding reactions with guanidine hydrochloride. This indicates that the reduction of the disulphide bond of the 33-kDa protein does not proceed through a species with a conformation very similar to that of the fully unfolded one, but through a species in which the disulphide bond is exposed by local fluctuations. This is different from the reduction mechanism for the C_L fragment in which the disulphide bond is reduced through a species with a completely unfolded conformation (see above). In the case of the 33-kDa protein, the disulphide bond is also located in the interior of the molecule but may not be completely buried, and is located in a part where local fluctuation occurs frequently.

D) Hydrogen-deuterium Exchange

When a protein is dissolved in heavy water, even hydrogen atoms in the interior of the protein molecule are exchanged with deuterium. Hydrogens attached to oxygen, nitrogen and sulphur are exchangeable with deuterium, whereas those attached to carbon are usually unexchangeable. Only the C-2 proton of the imidazole ring of histidine is exchangeable among the hydrogens attached to carbon under mild exchange conditions. Linderstrøm-Lang (1955) was the first to attempt a study of protein conformation using hydrogen-deuterium exchange. In his method, the hydrogens of a protein were first deuterated completely in heavy water and then the protein was lyophilized. This deuterated sample was dissolved in

water, and samples of a given volume of the solution were with-
drawn at appropriate intervals and lyophilized. By measuring the
densities of the water (a mixture of H_2O and ^2H_2O) recovered on
lyophilization with a density gradient tube at extremely high
precision, it was possible to follow the process of exchange of
deuterium with hydrogen.

Linderstrøm-Lang and Schellman (1959) proposed the concept
of "motility" to describe the conformation of protein molecules.
The protein molecule always fluctuates, and its conformation
ranges from the native structure to the randomly coiled state. The
conformation of the protein molecule between these extremes is
determined by the free energy of fluctuation of the molecule in the
environment. In its native environment, the average conformation
of the protein molecule is predominantly the native and compact
structure, and the results obtained by optical rotation or viscosity
provide information about the average conformation. Different
information is obtained by hydrogen-deuterium exchange however.
In order to undergo exchange with deuterium, the protein molecule
must "open", and expose exchangeable hydrogen atoms in the
molecule to solvent. The rate of hydrogen-deuterium exchange
within the protein molecule depends on both the rate of opening of
the molecule and the fraction of opened molecules present.

It is difficult to measure the exchange process with high
precision by the method of Linderstrøm-Lang. Furthermore, this
method measures not only the exchange of peptide NH protons but
also the exchangeable hydrogens of side chains. NMR and infrared
absorption spectroscopy, however, makes it possible to measure the
exchange of only the peptide NH protons. In the infrared absorption spectra of proteins, the amide II absorption band, which
corresponds to the N-H deformation frequency, shifts from 1,550
cm^{-1} to 1,450 cm^{-1} on deuteration. Thus the process of deuteration
of NH protons can be measured using the decrease in absorption at
1,550 cm^{-1}. NMR provides a powerful means of measuring the
exchange process, since a particular peptide NH proton or side
chain proton can be assigned unambiguously. The process of
exchange of the indole NH proton of Trp can also be followed by

measuring the change in ultraviolet absorption on deuteration, and on deuteration of the phenolic OH proton of Tyr, the fluorescence spectrum, but not the absorption spectrum, changes.

Hvidt and Nielsen (1966) proposed the following mechanism to explain hydrogen-deuterium exchange in proteins.

$$N(H) \underset{k_2}{\overset{k_1}{\rightleftharpoons}} O(H) \overset{k_3}{\longrightarrow} O(D) \underset{k_1}{\overset{k_2}{\rightleftharpoons}} N(D) \tag{6.5}$$

In this mechanism, N is the native protein molecule, in which the hydrogen under consideration is buried in the interior of the molecule and thus not accessible to solvent. O is any conformation of the protein molecule in which the hydrogen is accessible to solvent. k_1, k_2 and k_3 are the rate constants for the respective processes. The two-process model has generally been assumed to explain the exchange reaction of native proteins: exchange due to local fluctuations and exchange from the unfolded state after major cooperative unfolding (Woodward *et al.*, 1982).

In general, k_2' is much larger than k_1, and the exchange rate constant, k_{ex}, may be expressed by

$$k_{ex} = \frac{k_1}{k_2 + k_3} k_3. \tag{6.6}$$

When $k_3 \gg k_2$

$$k_{ex} = k_1 \tag{6.7}$$

When $k_3 \ll k_2$

$$k_{ex} = (k_1/k_2)k_3 = Kk_3, \tag{6.8}$$

where K is the equilibrium constant between N and O.

The chemical exchange of the peptide NH is catalyzed by acids and bases, and the chemical exchange rate constant, k_3, is given by

$$k_3 = k_H[H^+] + k_{OH}[OH^-], \tag{6.9}$$

where k_H and k_{OH} are the acid and base catalytic rate constants, respectively. A minimum exists in the plot of k_3 against pH where the two terms on the right of Eq. (6.9) are equal.

The exchange reaction of NH protons in a randomly-coiled polypeptide chain is affected not only by nearby charges but also by distant charges. The exchange rate for the NH protons of poly-L-lysine is affected greatly by the ionic strength of the medium, but that for the NH protons of poly-DL-alanine with no charges is not. The activation energy values for k_H and k_{OH} determined using poly-DL-alanine are 13 and 17 kcal/mol, respectively.

One of the most interesting aspects of hydrogen isotope exchange is to clarify whether the exchange proceeds through global unfolding or local fluctuations of the protein molecule under physiological conditions. The C_L fragment of the type-λ immunoglobulin light chain has two tryptophan residues at positions 150 and 187. Trp 150 is buried in the interior and Trp 187 lies on the surface of the molecule. Kawata *et al.* (1988) studied the hydrogen-deuterium exchange kinetics of the indole NH proton of Trp 150 at various p^2H values at 25°C by ^1H nuclear magnetic resonance. These two indole NH protons are distinguishable on the basis of NMR measurements. When the C_L fragment is exposed to ^2H_2O, the indole NH proton of Trp 187 is rapidly exchanged with deuterium, but the exchange kinetics for the indole NH proton of Trp 150 follow a first-order reaction. In Fig. 6-5, the exchange rate constants (k_{ex}) are plotted against p^2H. The exchange rate constant is relatively constant between p^2H 7 and 8, decreases below p^2H 7 and increases above p^2H 8. The C_L fragment is not unfolded above p^2H 8 (see Fig. 5-5), and the slope of the increase in the exchange rate is 1.1, which is very close to that for the base catalysis of the chemical exchange step of the indole NH protons of L-tryptophan derivatives. Therefore, we may assume that the increase in the exchange rate above p^2H 8 is due to the increase with pH in the chemical exchange rate of the indole NH proton of Trp 150 exposed by local fluctuations, and that the exchange is expressed by Eq. (6.8). When the contribution of this exchange through local fluctuations is subtracted from the observed exchange rates, the dotted line shown in Fig. 6-5 is obtained. The exchange rate levels off above pH 7 and the rate constant at this level was found to be 2×10^{-5} sec^{-1}. This value is very close to that (7×10^{-5} sec^{-1}) for the global unfolding

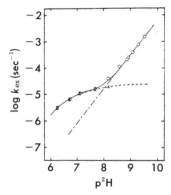

Fig. 6-5. Logarithmic plot of the exchange rate constants for the Trp 150 indole NH proton against p^2H at 25°C (○). The dashed-and-dotted line represents the contribution of the local fluctuations to the exchange, and the crosses and broken line represent that of the global fluctuations to the exchange. (Kawata, Y. *et al.* (1988) *Biochemistry, 27*, 346)

rate estimated from reduction of the disulphide bond of the C_L fragment under the same conditions (see above). Furthermore, the activation energy for the exchange process at pH 7.8 (27.7 kcal/mol) was the same as that for the process of unfolding by 2 M guanidine hydrochloride. These findings show that the exchange between pH 7 and 8 occurs through global unfolding of the protein molecule and is rate-limited by the unfolding even at 25°C, which is far below the transition temperature (60°C) of thermal unfolding of the C_L fragment. The same observations were also made for the C_L and V_L fragments of a type-$\varkappa$ immunoglobulin light chain (Kawata & Hamaguchi, 1990).

Hydrogen-deuterium exchange also occurs in protein crystals, and here we can determine directly which hydrogens are exchanged with deuterium. Diffraction with neutrons can distinguish hydrogen atoms and the deuterium isotope, because hydrogen has a negative scattering factor whereas deuterium has a large positive scattering factor. Therefore, exchange of a particular hydrogen atom in a protein with deuterium in the solvent may be followed using neutron diffraction. Kossiakoff (1982) studied hydrogen-deuterium exchange in trypsin crystals. A trypsin crystal was soaked in

^2H_2O solution at pH 7 and 20°C for about one year, and then the positions at which exchange with deuterium had occurred were identified by neutron diffraction. Trypsin has 215 exchangeable hydrogens on its amide groups. The degree of exchange of these amide hydrogens is shown in Fig. 6-6. Of the 215 exchangeable amide hydrogens, 146 (68%) were fully exchanged, 17 (8%) were partially exchanged and 52 (24%) were unexchanged. The most prominent feature was that unexchanged sites were clustered in regions corresponding to the β-sheet structure. Of the 52 unexchangeable hydrogens, 45 were found in β-structure. Similar observations were also made for hydrogen-deuterium exchange in a crystal of ribonuclease A (Wlodawer & Sjolin, 1982). Of the remaining 7

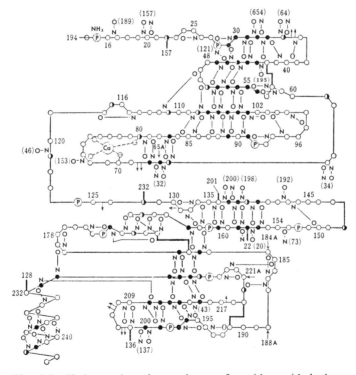

Fig. 6-6. Hydrogen-deuterium exchange of peptide amide hydrogens in trypsin crystals. ○, full exchange; ◑, partial exchange; ●, unexchanged; Ⓟ, proline; broad lines, S-S bond. (Kossiakoff, A.A. (1982) *Nature, 296*, 713)

unexchanged hydrogens, one was not involved in hydrogen bond formation, two were adjacent NH hydrogens in an α-helix, and two were formed with side chains. Almost all of the amide hydrogens in the two short helices of trypsin were exchanged except for the amide hydrogens of Ile 238 and Ile 242. This suggests that cooperative unfolding of the helix is not the cause of hydrogen-deuterium exchange; if there were cooperative unfolding of the whole helix, it would not be possible to explain why only the two NH hydrogens were unexchanged. Kossiakoff examined various factors such as the location of the sites of exchange and the temperature factors of amide hydrogens, which might determine the degree of exchange. However, it was not clear which factor was the predominant one affecting the exchange.

Binding of substrate or cofactors to enzymes greatly decreases the rates of exchange, and the mobile structures are frequently found to become more rigid upon substrate binding. The dynamic structures of proteins are clearly very important not only from the viewpoint of physicochemical properties, but also with regard to their biological function.

7
Protein Folding

In 1961, Anfinsen *et al.* demonstrated that the denatured ribonuclease A molecule, in which all the disulphide bonds are reduced, refolds into the original structure and regains its original activity upon oxidation in air. As criteria indicative of complete refolding, enzymatic activity, optical rotatory dispersion, viscosity, ultraviolet absorption, immunological reaction and peptide mapping were used. Upon oxidation of the completely reduced and unfolded ribonuclease A by atmospheric oxygen to form four disulphide bonds, the eight cysteine residues could theoretically have recombined at random in 105 different ways. When the first disulphide bond is formed from any one of the eight cysteine residues, there are seven ways of pairing, and when a second disulphide bond is formed from the remaining six cysteine residues, there are five ways of pairing, and so on. Thus, when four disulphide bonds are formed at random from eight cysteine residues, there are $7 \times 5 \times 3 \times 1 = 105$ different ways of pairing. In fact, however, only proteins with disulphide bonds between the cysteine residues identical to those of the original structure are formed preferentially from the completely reduced and unfolded protein by air oxidation under native conditions. This experiment carried out by Anfinsen established the principle that the conformation of the native protein molecule is in a state of minimum free energy, and that the linear sequence of the polypeptide chain contains all the information required to fold the chain into its native regular conformation.*

Genetic information determines the amino acid sequences of proteins, but is not required for formation of the three-dimensional

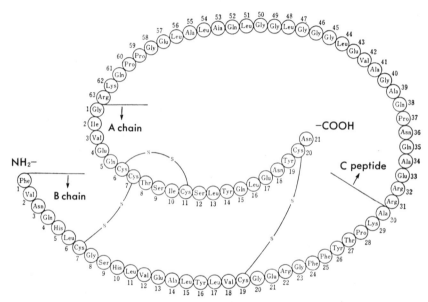

Fig. 7-1. The primary structure of proinsulin. (Chance, R.E. *et al.* (1968) *Science, 161*, 165)

---

*Correct formation of the disulphide bonds from completely reduced and unfolded proteins has been demonstrated for many other proteins since then. However, in the case of insulin, it was found that the recovery of activity was exceptionally poor when reduced insulin was reoxidized. This led to the discovery of proinsulin and other proproteins. Insulin consists of two chains, A and B, connected by two interchain disulphide bonds. Proinsulin consists of a single polypeptide chain (Fig. 7-1). *In vivo*, insulin is first synthesized as inactive proinsulin, and then formed by cleavage with a trypsin-like enzyme. Proinsulin may be recovered in high yield from the completely reduced and unfolded proprotein; active insulin may be formed from the recovered proinsulin on treatment with trypsin.

structure. Formation of the three-dimensional structure of proteins after the linear sequence has been synthesized is a spontaneous thermodynamic process in which the free energy decreases to a minimum. Recently, refolding of unfolded proteins has been actively studied, and in this chapter the mechanism of refolding is discussed.

7-1. MECHANISM OF UNFOLDING AND REFOLDING

When the unfolding conditions are removed, most unfolded small proteins refold to their native conformation. There have been many attempts to clarify the pathway of formation of the native conformation from the unfolded protein. As described in Section 5-2, however, unfolding and refolding can be approximated by a two-state transition, and it is difficult to detect any intermediate in the unfolding and refolding processes by equilibrium studies. Therefore, many studies have addressed the feasibility of detecting an intermediate kinetically. If a kinetic intermediate exists transiently in the unfolding and refolding processes, this would be expected to provide a clue to understanding the folding mechanism of the protein molecule.

In 1975, Garel and Baldwin and Brandts *et al.* studied the kinetics of unfolding and refolding of ribonuclease A by guanidine hydrochloride, acid and heat, and found that the kinetics follow the following three-species mechanism

$$D_S \underset{\longleftarrow}{\overset{\text{slow}}{\rightleftharpoons}} D_F \underset{\longleftarrow}{\overset{\text{fast}}{\rightleftharpoons}} N \cdot \qquad \text{(mechanism 1)}$$

In this mechanism, N is the native protein. D_F is not an intermediate in the unfolding and refolding process: D_F and D_S are both unfolded proteins and are indistinguishable in terms of their spectroscopic properties such as ultraviolet absorption and fluorescence. When ribonuclease A is unfolded, N is converted rapidly to D_F, which is then converted slowly to D_S. When the unfolded protein is refolded, D_F refolds to N rapidly, but refolding from D_S to N is rate-limited by the transition from D_S to D_F.

What is the difference between the rapid-folding species (D_F) and slow-folding species (D_S) of unfolded protein? In 1975, Brandts *et al.* interpreted the interconversion between the two forms (D_S and D_F) in the unfolded state in terms of the *cis-trans* isomerization of the peptide bond X-Pro, where X is any residue (including Pro). As described in Chapter 2, although the probability of a peptide bond occurring in the *cis* form is very small, the peptide bond, X-Pro, has a greater probability of occurring in the *cis* form. Therefore, an X-Pro bond fixed in either the *trans* or *cis* form in the native state can assume a different configuration by isomerization in the unfolded state. The configuration of an X-Pro bond in the slow-folding species (D_S) is different from that in the native state, and the configuration of the X-Pro bond in the rapid-folding species (D_F) is the same as that in the native protein. When a protein is unfolded, N is converted rapidly to D_F, which has the same X-Pro bond configuration as that in the native protein, and D_F is then slowly converted to D_S by isomerization of the X-Pro bond. In the refolding process, D_F can refold rapidly to N, but D_S can refold to N only after D_S converts to D_F by isomerization of the X-Pro bond.

The evidence that the slow kinetic phase is generated by proline imide isomerization has relied solely upon comparison between the folding kinetics of proteins and that of *cis-trans* isomerization of proline peptides. For the reaction $D_F \rightleftharpoons D_S$ of ribonuclease A, the equilibrium constant is 0.25, the enthalpy change is zero, the half-life time is 40 sec and the activation enthalpy is 20 kcal/mol. For the *cis-trans* isomerization reaction of proline peptides, the equilibrium constant is 0.1–1.0, the enthalpy change 0–1 kcal/mol, the half-life time 1,000 sec and the activation enthalpy 20 kcal/mol. Furthermore, both reactions are catalyzed by strong acid.

A more direct method for detecting the *cis* and *trans* isomers of specific proline residues is to use the isomer-specific proteolysis of proteolytic enzymes. Lin and Brandts (1983a, b, c, d, 1984, 1985) studied the isomer-specific proteolysis of proline-containing peptides by trypsin, chymotrypsin and prolidase. In general, the subsites of the active site for a proteolytic enzyme are represented by S_1, S_2, ⋯, S_1', S_2', ⋯ and the residues of a substrate corresponding to these subsites by P_1, P_2, ⋯, P_1', P_2', ⋯ (Fig. 7-2). It was found that trypsin can hydrolyze the Lys-X bond of -Lys-X-Pro- or the Arg-X bond of -Arg-X-Pro- only when X-Pro has the *trans* configuration (Pro is located at P_2'). When N-Cbz-Gly-Pro-Arg-*p*-nitroanilide is used as a substrate, Pro is located at P_2. In this case, the Arg-*p*-nitroanilide bond is also hydrolyzed even when the Gly-Pro has the *cis* configuration. Chymotrypsin can also hydrolyze the Tyr-Ala bond of Ala-Ala-Ala-Pro-Tyr-Ala-Ala-Ala even when the Ala-Pro bond is in the *cis* configuration (Pro is located at P_2). Thus the isomer specificity depends upon the position of the bond subjected to hydrolysis relative to the position of the X-Pro bond.

Utilizing isomer-specific proteolysis, Lin and Brandts attempted to determine whether the $D_S \rightleftharpoons D_F$ reaction is generated by the *cis-trans* isomerization reaction of X-Pro bonds. Ribonuclease A has four Pro residues at positions 42, 93, 114 and 117 with *trans*, *cis*, *cis* and *trans* configurations, respectively. The refolding kinetics of ribonuclease A consists of a fast phase (20%) and a slow phase (80%). The slow phase is further divided into three phases: a fast phase (XY, 50%), a slow phase (CT, 25-30%) and a slower phase (ct, 5%). Ribonuclease A contains the sequence -Lys-Tyr-Pro-.[91][92][93] Utilizing the isomer specificity of trypsin, which hydrolyzes the Lys-Tyr bond only when the Tyr-Pro bond is in the *trans* configuration, it was found that the *cis-trans* isomerization of Tyr-Pro is responsible for the appearance of the CT phase. Ribonuclease A also contains the sequence -Gly-Asn-Pro-Tyr-Val-Pro-.[114][117] Using a proline-specific endopeptidase that acts only when the Asn-Pro[113][114]

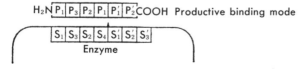

Fig. 7-2. Mode of binding of peptide substrate with proteolytic enzyme. ▲, the site of cleavage.

bond is in the *trans* configuration, and aminopeptidase P, which acts only when the Val-Pro bond is in the *trans* configuration, it was found that Pro 114 is responsible for the appearance of the ct phase, and that Pro 117 is not involved in any of the slow phases. It was also found that the XY phase is not related to the isomerization reaction of any of the Pro residues, so that other factors seem to be involved in the appearance of this phase.

Saccharomyces cerevisiae iso-1-cytochrome *c* has Pro at position 71. Comparison of the unfolding and refolding kinetics of the wild-type cytochrome *c* with those of a mutant protein in which Pro 71 is replaced by Val, Thr or Ile showed that Pro 71 is not responsible for the appearance of the slow phase (Ramdas & Nall, 1986). Thioredoxin has a sequence -Ile-Pro- with a *cis* configuration. A mutant protein in which Pro 76 is replaced by Ala showed no slow phase (Kelly & Richards, 1987). This suggested that the isomerization of Pro 76 is responsible for the appearance of the slow phase.

Taken together, these findings indicate that the *cis-trans* isomerization reaction of X-Pro is not always responsible for the appearance of slow phases in unfolding and refolding kinetics. Recently, peptidyl-prolyl *cis-trans* isomerase, which catalyzes the *cis-trans* isomerization of the X-Pro bond, has been discovered in mammalian tissues and purified from porcine kidney (Fischer *et al.*, 1984). This enzyme was found to accelerate the slow unfolding reactions of ribonuclease T1 and immunoglobulin light chain (Lang *et al.*, 1987; Fischer *et al.*, 1989; Kiefhaber *et al.*, 1990a, b), indicating that *cis-trans* isomerization reactions of X-Pro bonds are responsible for the appearance of the slow phase of the refolding processes of these proteins.

The facts described above show that it is difficult to trap intermediates in the process of refolding from an unfolded protein either by equilibrium studies or kinetic studies. Because of the cooperative character of protein folding, stable intermediates do not accumulate in the refolding process. There are two models, the framework model and the jigsaw puzzle model, for the mechanism of protein folding. In the framework model, it is postulated that secondary structures play an important role in determining the folding pathway. As described by Baldwin (1989): "Folding begins with the formation of transient secondary structures that are stabilized by packing against each other. The folding is a hierarchical process in which simple structures are formed first and then interact to give more complex structures". In the jigsaw puzzle model, on the other hand, it is postulated that there is no unique folding pathway and that there are many ways of assembling the "jigsaw puzzle" (Harrison & Durbin, 1985). To explain the difference between the two models, let us take the example of how elderly and young Japanese write "kanji" (Chinese characters). The elderly observe the rules which stipulate the order of strokes, whereas the young write the same kanji ignoring such rules. However, the same character written by the two generations looks exactly the same. The process of construction is different, but the result is the same. This can be applied to the case of the framework model and the jigsaw puzzle model, the former representing the kanji written by elderly Japanese and the latter, that written by the young.

The following recent observations support the framework model.

1) As described in Section 2-4, S-peptide and C-peptide isolated from ribonuclease A form an α-helix. This suggests that when ribonuclease A is refolded from its unfolded state, an α-helix is formed in the peptide segment corresponding to the S-peptide or C-peptide in the early stage of refolding.

2) Recently ^{1}H-NMR measurements combined with hydrogen-deuterium exchange have been used to characterize the folding intermediates of ribonuclease A (Udgaonkar & Baldwin, 1988) and cytochrome c (Roder et $al.$, 1988). The proteins are first unfolded in a concentrated guanidine hydrochloride solution in ^{2}H$_2$O. Refolding is initiated by diluting this solution with ^{2}H$_2$O. When secondary and tertiary structures are partially formed at an early stage of refolding, some deuterium atoms are protected by these regular structures whereas others are not protected and are exposed to the solvent. Immediately after dilution, a labelling pulse is

initiated by dilution into excess H_2O, and the pH is kept high in order to accelerate the hydrogen exchange reaction. If an intermediate with partial secondary and tertiary structure is formed, deuterium atoms which are protected by these structures are not exchanged with hydrogen, whereas deuterium atoms which are exposed to the solvent are exchanged rapidly. Finally, the labelling pulse is terminated by lowering the pH so that it approximates the value at which hydrogen exchange rates are slowest. Using NMR, deuterium atoms which either have or have not been exchanged are identified, and thus the structure of the intermediate trapped in the refolding process can be determined. The results thus obtained favour the framework model.

3) Stopped-flow CD measurements have shown that secondary structure is formed at an early stage of the refolding of hen egg-white lysozyme (Kuwajima *et al.*, 1985).

Kawata and Hamaguchi (1991) prepared the C_L fragment of a type-κ immunoglobulin light chain in which the C-terminal cysteine residue was modified with N-(iodoacetyl)-N'-(5-sulpho-1-naphthyl) ethylene diamine (C_L-AEDANS fragment). This C_L fragment has only one tryptophan residue at position 148. The compactness of the modified fragment was measured in the early stages of refolding from 4 M guanidine hydrochloride by fluorescence energy transfer from Trp 148 to the AEDANS group. In 4 M guanidine hydrochloride, the distance between the donor and the acceptor is larger, and the efficiency of the energy transfer lower. The distance between Trp 148 and the AEDANS group for the intact protein estimated using the energy-transfer data was in good agreement with that obtained by X-ray crystallographic analysis. The kinetics of unfolding and refolding of the modified fragment were then measured by the use of fluorescence energy transfer, tryptophyl fluorescence and CD at 218 nm. These three methods gave the same kinetic unfolding pattern. However, the refolding kinetics measured by fluorescence energy transfer were different from those measured by tryptophyl fluorescence and CD, which gave the same kinetic pattern. In addition to the two phases observed using tryptophyl fluorescence or CD, a very much faster phase was detected by fluorescence energy transfer (Fig. 7-3). The energy-transfer efficiency reached the same level as that of the intact protein at a very early stage of refolding. Double-jump experiments also gave the same result. These findings indicate that a structure as compact as that of the native protein is formed immediately after refolding, and that the compact molecule then converts slowly to the native protein by rearrangements of various groups, probably involving *cis-trans* isomerization of the prolyl residue. Kawata and Hamaguchi (1991) proposed the following mechanism to explain the unfolding and refolding kinetics of the C_L fragment at guanidine hydrochloride concentrations below the transition zone

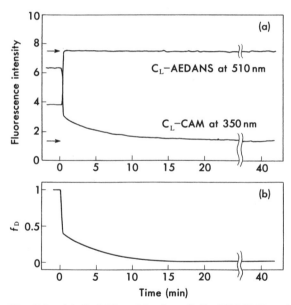

Fig. 7-3. (a) Refolding kinetics of C_L-AEDANS and unmodified fragments obtained by monitoring the change in fluorescence energy transfer efficiency at 510 nm and the change in tryptophyl fluorescence at 350 nm, respectively. (b) Refolding kinetics of C_L-AEDANS and unmodified C_L fragments obtained by monitoring the change in ellipticity at 218 nm. The refolding kinetics for the two fragments were identical. (Kawata, Y. & Hamaguchi, K. (1991) *Biochemistry, 30,* 4367)

$$
\begin{array}{ccc}
& \text{slow} & \\
U_1^E & \rightleftharpoons & U_2^E \\
\text{very} \Big\updownarrow & & \Big\updownarrow \text{ very} \\
\text{fast} & & \text{fast} \\
U_1^C & \rightleftharpoons & U_2^C \rightleftharpoons N. \\
& \text{slow} & \text{fast}
\end{array}
\qquad \text{(mechanism 2)}
$$

In this mechanism, N is the native protein, U_1 and U_2 are the slow-folding and fast-folding species of unfolded protein, respectively, U^E is the extended unfolded molecule or the unfolded molecule with a large volume, and U^C is the molecule whose conformation is disordered, but nearly as compact as the native one. The rates for the reactions $U^E \rightleftharpoons U^C$, $U_2^C \rightleftharpoons N$ and $U_1^C \rightleftharpoons U_2^C$ are in the order of milliseconds, seconds, and minutes, respectively. Although they did not measure directly the compactness of the protein molecule, but determined the absorption at 292 nm, Kuwajima *et al.* (1989) also proposed a similar mechanism for the folding of α-lactalbumin.

As described above, the three-species mechanism (mechanism 1) has been proposed for the unfolding and refolding of proteins. This model describes only the presence of two unfolded species in the unfolded state. Mechanism 2 describes further the conformation of the protein molecule immediately after transfer from unfolding conditions to refolding conditions. The appearance of such a compact structure in the refolding process has been suggested for some proteins. Although the secondary structure of the U^C molecule in the C_L-AEDANS fragment has not been measured directly, CD measurements have suggested that the U^C molecule has about 50% of the secondary structure of the folded one. These characteristics of the U^C molecule are very similar to those of the molten globule state. As described above, recent studies have indicated that the molten globule state, which is compact and has secondary structure formed at least partly in the same segments as those existing in the native protein, appears at an early stage of refolding.

7-2. FORMATION OF DISULPHIDE BONDS

There have been many investigations of the formation of

intrachain disulphide bonds from disulphide-reduced proteins, and two mechanisms have been considered for the folding of reduced and unfolded proteins. One assumes that there is a limited number of folding pathways. The other assumes that many incorrectly paired disulphide bonds are formed during the early stage of reoxidation, which then shuffle into the correct pairing. Assuming that each pair of cysteine residues that comes into regional proximity can always form a disulphide bond, and that disulphide formation is correlated with the conformational transitions of the polypeptide chain, Creighton (1975) studied the pathway of disulphide bond formation in bovine pancreatic trypsin inhibitor (BPTI). BPTI consists of 58 residues and has three disulphide bonds (Cys 14–78, Cys 5–55 and Cys 30–51) (see Fig. 2-24). Disulphide bond formation was initiated by adding oxidized glutathione to BPTI in which all of the three disulphide bonds were reduced. The reaction was stopped by addition of monoiodoacetic acid at appropriate time intervals, and the pairs of cysteine residues which had formed disulphide bonds were determined. The results are shown in Fig. 7-4. It is evident that there is a pathway of disulphide bond formation, but that correct disulphide bonds are not formed successively. Two single-disulphide intermediates, one with a disulphide bond between Cys 30 and Cys 51 (designated (30–51)) and the other (5–30) were formed predominantly in the early stage. Second disulphide bonds were formed only after (30–51) formation, and (30–51, 5–14), (30–51, 5–38), and (30–51, 14–38) were then formed. The disulphide bonds 5–14 and 5–38 are not present in the native BPTI; intermediate (30–51, 14–38), however, has two native disulphide bonds. None of these three two-disulphide species can form a third disulphide bond directly, but (30–51, 5–55) was formed by rearranging the second disulphide bonds. This species has two native disulphide bonds and a native-like conformation. Finally, the third disulphide bond was formed between Cys 14 and Cys 38 to complete the native disulphide bond formation.

Two points should be considered in interpreting the results of Fig. 7-4. (1) In order to understand the mechanism of refolding, it may be necessary to determine the conformation of the reduced

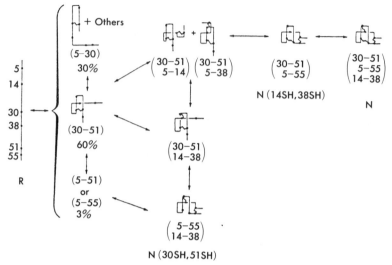

Fig. 7-4. The pathway of disulphide bond formation in BPTI. (Creighton, T.E. & Goldenberg, D.P. (1984) *J. Mol. Biol.*, *179*, 497)

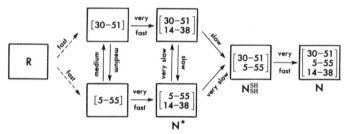

Fig. 7-5. The pathway of disulphide bond formation in BPTI determined by Weissman and Kim (1991).

protein from which the native conformation begins to emerge in aqueous solution. If reduced BPTI has some regular conformation in aqueous solution, this may determine the pathway of disulphide bond formation. As described in Section 5-5, no conformational change occurs upon reduction of the disulphide bond of the C_L fragment, and thus the two SH groups are located in close proximity. Reduced hen egg-white lysozyme (White, 1976) and reduced Taka-amylase A (Takagi & Isemura, 1966) have been shown to contain some secondary structure in aqueous solution. (2) It was found that a considerable amount of the species (5–55, 14–38) accumulated in the refolding pathway of reduced BPTI (Fig. 7-3). Although this species has native disulphides 5–55 and 14–38, it cannot form the third native disulphide bond between Cys 30 and Cys 51. This suggests that the conformation of (5–55, 14–38) is very stable, and that no further reoxidation reaction proceeds, because Cys 30 and Cys 51 are buried in the interior of the (5–55, 14–38) molecule.*

The correlation between disulphide bond formation and protein folding will only be understood unambiguously by studying the regeneration of a protein with only one intrachain disulphide bond. Each domain of the immunoglobulin molecule has only one intrachain disulphide bond buried in the interior of the hydrophobic region between two β-sheets. The C_L fragment whose intrachain disulphide bond is reduced has a conformation very similar to that of the intact C_L fragment, and the two cysteine residues are buried in the interior of the protein molecule. Goto and Hamaguchi (1981) studied disulphide bond formation in the reduced C_L fragment by oxidized glutathione in the absence and presence of 8 M urea. Oxidation of the reduced C_L fragment (C_{LSH}^{SH}) with glutathione (GSSG) yielded four species, C_{LSH}^{SH}, C_{LSH}^{SSG} (C_L fragment in which the intrachain disulphide bond is reduced, and one of the two cysteine residues forms a mixed disulphide), C_{LSSG}^{SSG} (C_L fragment in which the intrachain disulphide bond is reduced, and the two cysteine residues form a mixed disulphide) and C_{LS}^{S} (C_L fragment with an intact intrachain disulphide bond). These species were trapped by addition of iodoacetamide and identified by electrophoresis in polyacrylamide gel. Figure 7-6 shows the changes with time of the concentrations of species trapped with iodoacetamide during the reaction of C_{LSH}^{SH} with GSSG in the

---

*Very recently, Weissman and Kim (1991) reinvestigated the pathway of folding from reduced BPTI, and the folding pathway they obtained (Fig. 7-5) is quite different from that shown in Fig. 7-4. All well-populated intermediates contain only native disulphide bonds and non-native intermediates are not populated significantly.

190

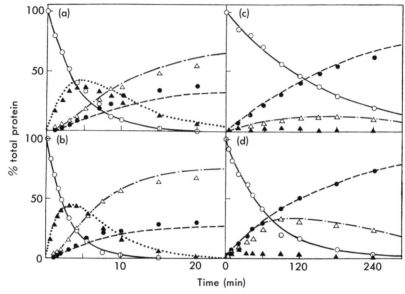

Fig. 7-6. Kinetics of the reaction of the reduced C_L fragment with (a) 2mM and (b) 3mM GSSG in the presence of 8 M urea and with (c) 3 mM and (d) 9 mM GSSG in the absence of urea at pH 8.1 and 25°C. C_{LSH}^{SH} (○), C_{LSH}^{SSG} (▲), C_{LSSG}^{SSG} (△), C_{LS}^{S} (●). (Goto, Y. & Hamaguchi, K. (1981) *J. Mol. Biol., 146*, 321)

absence and presence of 8 M urea. The following two distinct differences were observed between the reaction in 8 M urea and that in the absence of urea. (1) The reaction is much slower but the yield of C_{LS}^{S} is much higher in the absence of urea than in its presence, when compared at the same concentration of GSSG. (2) Compared with the reaction of C_{LSH}^{SH} with GSSG in 8 M urea, the accumulation of C_{LSH}^{SSG} is much lower and the yield of C_{LS}^{S} much higher in the absence of urea. The slowness of the reaction in the absence of urea is due to the two cysteine residues of the reduced C_L fragment being buried in the interior of the molecule, and because the oxidized glutathione is capable of reacting with the thiols only in the opened form of the protein molecule. The high yield of C_{LS}^{S} is due to the cysteine thiol and the mixed disulphide in the intermediate (C_{LSH}^{SSG}) being located appropriately nearby. In the presence of 8 M urea, the cysteinyl thiol and mixed disulphide in the intermediate are far apart and the process of disulphide formation is a statistically controlled event.

The thiol-disulphide interchange reactions of the C_L fragment with glutathione in the presence and absence of 8 M urea are shown in Figs. 7-7 and 7-8, respectively; the rate constants in these figures are summarized in Table 7-1.

The above results for the disulphide formation of the C_L fragment suggest that when cysteine residues are buried in an

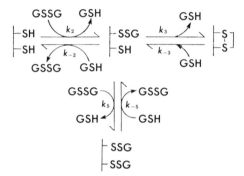

Fig. 7-7. Thiol-disulphide interchange reaction of the C_L fragment with gluta-thione in the presence of 8 M urea. k_2, k_{-2}, k_{-3}, k_5 and k_{-5} are the second-order rate constants for the thiol-disulphide interchange reactions and k_3 is the first-order rate constant for the formation of the intrachain disulphide bond. (Goto, Y. & Hamaguchi, K. (1981) *J. Mol. Biol., 146*, 321)

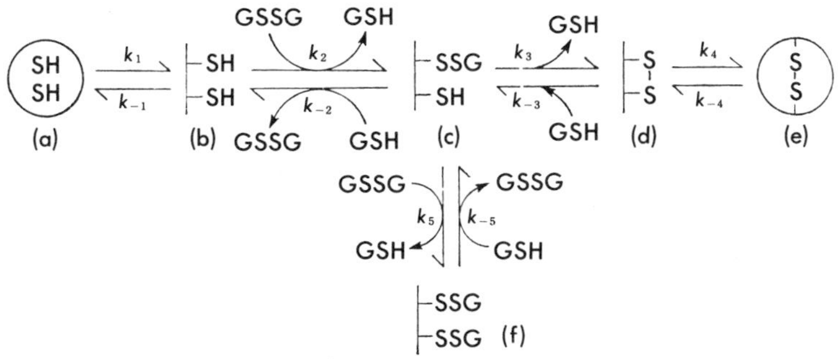

Fig. 7-8. Thiol-disulphide interchange reaction of the C_L fragment with gluta-thione in the absence of urea. k_2, k_{-2}, k_{-3}, k_5 and k_{-5} are the second-order rate constants for the thiol-disulphide interchange reactions and the others are the first-order rate constants for the intramolecular reactions. (Goto, Y. & Hamaguchi, K. (1981) *J. Mol. Biol., 146*, 321)

TABLE 7-1

Values of the Rate Constants for Reaction of C_L Fragment with Glutathione (pH 8.1, 25°C)

	k_1 (sec^{-1})	k_{-1} (sec^{-1})	k_2 (sec^{-1} M^{-1})	k_{-2} (sec^{-1} M^{-1})	k_3 (sec^{-1})
In absence of urea	6.3×10^{-4}	2.8×10^{-2}	2.0	1.7	8.3×10^{-3}
					6.2×10^{-3}
In 8 M urea	—	—	2.0	6.7×10^{-1}	1×10^{-3}

	k_{-3} (sec^{-1} M^{-1})	k_4 (sec^{-1})	k_{-4} (sec^{-1} M^{-1})	k_5 (sec^{-1} M^{-1})	k_{-5} (sec^{-1} M^{-1})
In absence of urea	(2.1)	Very fast	Very slow	1.0	3.3
In 8 M urea	8.3×10^{-1}	—	—	1.0	1.3

Goto, Y. & Hamaguchi, K. (1981) *J. Mol. Biol., 146*, 321.

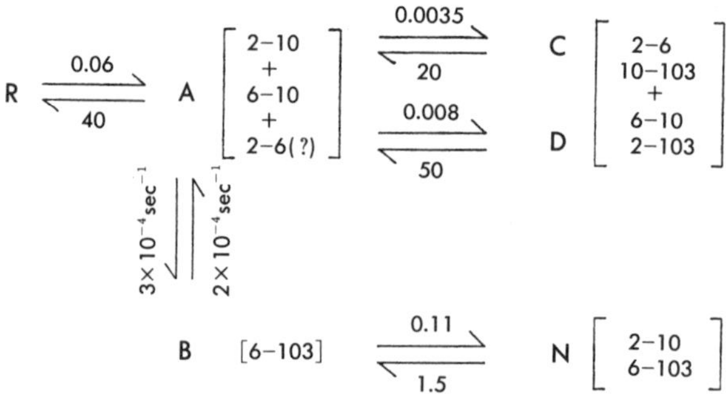

Fig. 7-9. The pathway of disulphide bond formation in ribonuclease T1. (Pace, C.N. & Creighton, T.E. (1986) *J. Mol. Biol., 188*, 477)

intermediate formed during the course of reoxidation of reduced BPTI, it is difficult to form a disulphide bond even if they are close enough to do so, because they are not accessible to solvent. Therefore, it may be important to study the conformations of the intermediates.

The pathway of disulphide bond formation from reduced ribonuclease T1 was studied by Pace and Creighton (1986) (Fig. 7-9). This protein has two disulphide bonds, between Cys 2 and Cys 10 and between Cys 6 and Cys 103. Three single-disulphide intermediates ((2–10), (6–10) and (2–6)) (A) were formed in the early stage. Intermediate B (6–103) was not formed directly from reduced ribonuclease T1 (R) but through formation of the A intermediates. The native protein (N) was not formed from two-disulphide intermediates (C and D) but from intermediate B. In this case, it may also be important to examine the conformations of the intermediates.

7-3. PROTEIN FOLDING *IN VIVO*

Although many studies have been carried out on *in vitro* folding of proteins, the mechanism of folding *in vivo* is more complex and not clearly understood. In addition to the problem of which factors are involved in the maturation and folding of nascent polypeptide chains *in vivo*, it is not clear whether the folding of proteins is initiated sequentially from the N-terminus (co-translational folding), or after completion of the polypeptide chain (post-translational folding). Furthermore, it is not clear whether disulphide bonds are formed co-translationally *in vivo* (Jaenicke, 1987; Wright *et al.*, 1988; Baldwin, 1989; Fischer & Schmid, 1990).

Phillips (1966) was the first to propose the idea that folding of the hen egg-white lysozyme molecule is initiated before the poly-peptide chain has been completely synthesized (Fig. 7-10). How-ever, his idea has not received much attention because it has been found that fragments isolated from proteins by chemical or en-zymatic treatment do not adopt the same conformations as those in the protein molecule. For instance, a fragment corresponding to the first 126 residues isolated from *Staphylococcus* nuclease, which consists of 149 amino acid residues, was found not to assume any regular conformation. Although des-(121–124)-ribonuclease A has a conformation very similar to that of the intact protein, products obtained by air-oxidation of reduced des-(121–124)-ribonuclease A contain randomly paired disulphide bonds. Although these find-ings do not seem to favour co-translational folding, there are many others which do. For instance, the product obtained by air-oxidation of reduced des-(128, 129)-hen egg-white lysozyme was found to be identical to the modified protein before reduction. This

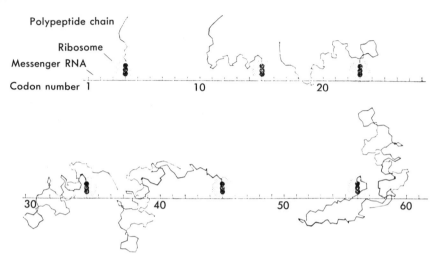

Fig. 7-10.   Folding of the lysozyme molecule *in vivo*. (Phillips, D.C. (1966) *Sci. Am., 215,* 78)

showed that the COOH-terminal dipeptide is not required for the native-like folding of the lysozyme molecule. This result is different from that of des-(121–124)-ribonuclease A described above. Proteins with no disulphide bonds fold extremely rapidly *in vitro*. For instance, the conformation of nuclease or aldolase is completed within 1 sec and that of myoglobin within 10 sec. This is at least 10 times faster than the time required for the biosynthesis of the complete linear sequence. Also, as described in Section 2-4, the C-peptide or S-peptide isolated from ribonuclease A forms an α-helix. This is consistent with initiation of folding of the protein from the N-terminus, apparently supporting Phillips' proposal.

Ishiwata *et al.* (1991) prepared C_L fragments with different carboxyl terminals, C_L (109–211) (fragment corresponding to sequence 109 to 211), C_L (109–207), and C_L (109–200), by limited proteolysis of the constant fragment (109–214) of a type-λ immunoglobulin light chain, and studied their conformations and stabilities, and the formation of the disulphide bond from the reduced fragments. It was found that removal of seven or more residues from the C-terminal end destabilized the C_L fragment. As described previously, the two cysteine residues of the reduced C_L

(109–214) fragment are buried in the interior of the molecule, and thus a long time is required for formation of the disulphide bond by oxidized glutathione. This suggests that the disulphide bond is formed before the completion of polypeptide synthesis. On the other hand, the conformation of the reduced C_L (109–207) fragment is disordered, and the two SH groups are exposed to solvent and can easily react with an oxidizing agent to form the disulphide bond. It was found that formation of the disulphide bond from the reduced C_L (109–207) fragment is about seven times faster than that from the C_L (109–214) fragment, suggesting that the disulphide bond is formed before complete folding of the C_L fragment. The work of Bergman and Kuehl (1979) has shown that the intrachain disulphide bonds of the MPC 11 light chain are also formed rapidly before completion of the primary structure.

Recently it has been shown that at least three proteins are involved in the folding of proteins *in vivo*.

1) Protein disulphide isomerase

This enzyme is found abundantly in the endoplasmic reticulum. Protein disulphide isomerase accelerates the rate of slow steps in the formation and breakage of disulphide bonds in bovine pancreatic trypsin inhibitor (Creighton *et al.*, 1990). Bulleid and Freedman (1988) prepared dog pancreas microsomes deficient in soluble luminal proteins including protein disulphide isomerase. These microsomal preparations were still able to translocate and process proteins synthesized *in vitro* but were defective in the formation of the correct disulphide bonds. This study provided conclusive evidence for the important role of protein disulphide isomerase in protein folding *in vivo*.

2) Peptidyl-prolyl cis-trans isomerase

As described previously, isomerization of the X-Pro peptide bond is one of the factors that slows down the protein folding reaction *in vitro*. Fischer *et al.* (1984) have found an enzyme that accelerates the *cis-trans* isomerization of X-Pro peptide bonds in short oligopeptides. The major binding proteins for the immunosuppressant FK506 and cyclophilin, which belong to a class of proteins termed immunophilins, also have peptidyl-prolyl *cis-trans* isomerase activity, although they are unrelated at the primary sequence level (Takahashi *et al.*, 1989; Fischer *et al.*, 1989; Moore *et al.*, 1991).

Although proline isomerization is not the only cause of slow folding of proteins *in vitro*, folding of some proteins is accelerated by peptidyl-prolyl *cis-trans* isomerase. For instance, slow refolding reactions of immunoglobulin light chain, ribonuclease A and ribonuclease T1 are accelerated by this isomerase. In particular, the slow step of ribonuclease T1 is accelerated more than 100-fold by its presence (Fischer *et al.*, 1989; Kiefhaber *et al.*, 1990a, b). On the other hand, refolding of chymotrypsinogen and thioredoxin was found not to be accelerated by the addition of this enzyme, although proline isomerization is involved in the refolding of these proteins.

The role of peptidyl-prolyl *cis-trans* isomerase in protein folding *in vivo* is not clear at present.

3) Molecular chaperones

The cellular proteins that are necessary for the correct folding of monomeric proteins and for the assembly of oligomeric proteins are called molecular chaperones. Such molecular chaperones are localized in *Escherichia coli* (the GroEL and GroES proteins), the chloroplasts of higher plants (the Rubisco subunit-binding proteins) and the mitochondria of *Saccharomyces cerevisiae* and *Neurospora crassa* (hsp (heat shock protein) 60). GroEL is known as cpn (chaperonin) 60 or hsp60 and GroES is known as cpn10 or hsp10. GroEL is a 14-*mer* consisting of identical subunits with 549 residues each. GroES is a 7-*mer* consisting of identical subunits with 97 residues each. GroEL has weak potassium-dependent ATPase activity.

Rubisco (ribulose biphosphate carboxylase) consists of eight large subunits and eight small subunits (expressed as L_8S_8). Rubisco subunit binding protein is indispensable for the correct assembly of L_8S_8, and assists in the oligomeric formation by binding to large subunits, but is not involved in the final product (Goloubinoff *et al.*, 1989). Rubisco subunit-binding protein and *E. coli* GroE protein are similar structurally and functionally.

The denaturation of citrate synthase, which is a dimeric protein, is irreversible *in vitro* because the refolding polypeptide chains aggregate rapidly, but addition of GroE and MgATP increases the yields of reactivated citrate synthase (Buchner *et al.*, 1991). GroE inhibits the aggregation reactions and facilitates the refolding of citrate synthase. GroEL forms a complex with unfolded or partially folded citrate synthase molecules and aggregation is protected by formation of this complex. Addition of GroES and ATP hydrolysis is necessary to release the polypeptide chain bound to GroEL and to allow further folding to its final state. GroE does not increase the rate of refolding of citrate synthase, however, and this is different from protein disulphide isomerase and prolyl isomerase, which accelerate the rate of folding.

Chaperonin-mediated folding of two monomeric enzymes, dihydrofolate reductase and rhodanese, has been studied *in vitro* by Martin *et al.* (1991). The GroEL protein stabilizes the protein in a condition similar to the molten globule state, and to gain the final regular conformation, hydrolysis of ATP and GroES is required.

For folding of large proteins and oligomeric proteins *in vivo*, at least the above three types of protein are apparently required; the information within the primary structure alone seems to be insufficient.

8

An Example of Enzyme Catalysis: Lysozyme

Hen egg-white lysozyme is one of the enzymes which has been studied most extensively with regard to its structure-function relationships. The complete amino acid sequence of hen egg-white lysozyme was determined independently by Jollès and his co-workers and by Canfield in 1963. Hen lysozyme consists of a single polypeptide chain of 129 amino acid residues, cross-linked by four disulphide bonds. Lysozyme is a basic protein with an isoelectric point of about 11. The physical and chemical methods which are generally employed for studying protein structure have all been tested for their validity using lysozyme as a representative protein, since it is a comparatively simple molecule of low molecular weight.

Lysozyme catalyzes the hydrolysis of cell-wall mucopolysaccharides comprised of alternate N-acetylglucosamine (GlcNAc) and N-acetylmuramic acid (MurNAc) residues joined by β-(1→4)-glycosidic linkages. The cell walls also include short polypeptide chains made up of both D- and L-amino acids, which are attached to the lactyl chains of some MurNAc residues (Fig. 8-1). Against the cell walls, lysozyme acts as an N-acetylmuramidase (N-acetyl-

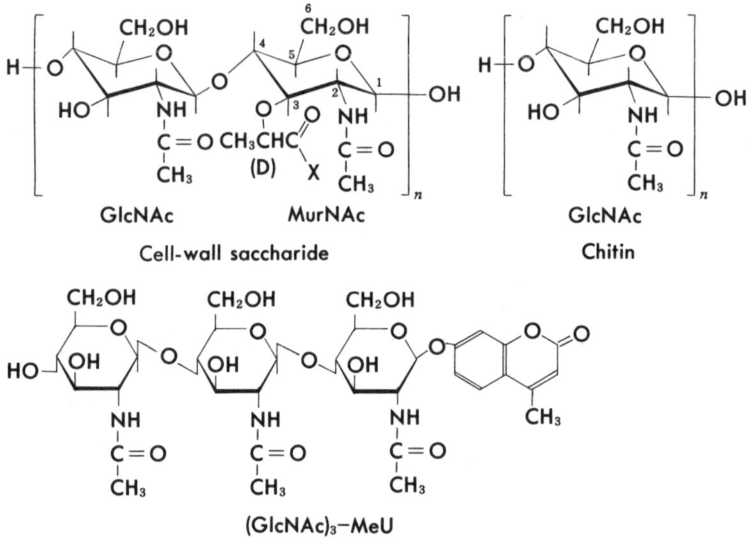

Fig. 8-1. Structures of cell-wall saccharide, chitin and 4-methylumbelliferyl-β-glycoside of GlcNAc trimer.

muramide glyconohydrolase), cleaving only the glycosidic bonds of MurNAc residues. Lysozyme also degrades chitin, the linear β-(1→4)-linked polymer of GlcNAc (Fig. 8-1), in this case acting as a chitinase (N-acetyl-glucosaminidase). Lysozyme activity has been measured by turbidimetry using bacterial cell walls. However, this lytic activity changes with ionic strength or pH in a complicated manner, due to the electrostatic interactions between the positively charged lysozyme molecule and negatively charged cell walls. Lysozyme activity can be readily measured by viscometry using a neutral soluble chitin derivative, glycol chitin (Hamaguchi & Funatsu, 1959). The activity-pH curves for hen egg-white, turkey egg-white and human lysozymes using glycol chitin as a substrate are shown in Fig. 8-2. The optimum pH is 4.8 for hen and turkey lysozymes and 5.1 for human lysozyme. Lysozyme exhibits trans-glycosylation activity in addition to hydrolytic activity. This trans-glycosylation activity complicates the study of the enzymatic properties of lysozyme.

The first report of an X-ray crystallographic study of hen

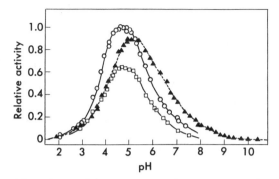

Fig. 8-2. Activity-pH curves of hen egg-white lysozyme (○), turkey egg-white lysozyme (□) and human lysozyme (▲). 25°C, 0.1 ionic strength. Substrate: glycol chitin.

egg-white lysozyme, by the Phillips group in England, appeared in 1965 (Blake *et al.*, 1965). Information on the more detailed structure and mode of substrate binding followed in 1967 (Blake *et al.*, 1967a, b) (see Figs. 2-20 and 2-21). Thus lysozyme became the first enzyme for which the enzymatic activity was understood on the basis of the three-dimensional structure of the molecule. In order to understand the enzymatic action in solution on the basis of the enzyme structure, it is necessary to know the physical and chemical properties of the amino acid residues located in the active site and the mode of interactions with substrate. This chapter describes how the hydrolytic reaction catalyzed by lysozyme can be explained on the basis of the ionization behaviour of the catalytic groups and of the ionizable groups in the active site and the pH dependence of the binding constants of substrate analogues.

8-1. STRUCTURE OF THE ACTIVE SITE

The lysozyme molecule is roughly ellipsoidal ($45 \times 30 \times 30$ Å) with a deep cleft. Substrate binds to this cleft and the complete binding site can accommodate six GlcNAc residues. The subsites for the sugar residues are denoted A, B, C, D, E and F, respectively (counting from the residue at the non-reducing terminus).

Crystallographic studies on the hen lysozyme-GlcNAc complex in the tetragonal form indicated that the two anomers, α- and β-GlcNAc, are bound to subsite C in similar but distinct ways. In both complexes, the N-acetyl moiety is coordinated near the ring of Trp 108 in the same configuration by hydrogen bonds between the NH and CO of the N-acetyl group and the main chain CO and NH groups of residues 107 and 59, respectively. However, O(6) and O(3) of β-GlcNAc are hydrogen-bonded to the side-chain NH groups of Trp 62 and 63, respectively, whereas β-GlcNAc is coordinated by O(1) to the backbone NH of residue 109. On binding of GlcNAc, the substrate binding cleft is narrowed, the movement of Trp 62 being most evident.

The structure of the GlcNAc-hen lysozyme complex in the triclinic form was studied by Kurachi *et al.* (1976) at 2 Å resolution. It was found that only the β-anomer is bound at subsite C in the triclinic crystals. On binding of β-GlcNAc, Trp 62 moves towards the bound GlcNAc by 0.5 Å and the carbonyl group of Ala 107 also moves towards the bound GlcNAc by about 0.2 Å. The latter movement was not observed clearly in the tetragonal crystals. The four hydrogen bonds between the bound GlcNAc and lysozyme molecule are very similar to those found in the tetragonal crystals.

The structure of the (GlcNAc)$_2$-hen lysozyme complex in the tetragonal form was studied at 6 Å resolution by Johnson and Phillips (1965). It was suggested that (GlcNAc)$_2$ binds in two different ways. Kurachi *et al.* (1976) studied the complex at 2 Å resolution and showed that (GlcNAc)$_2$ is bound only at subsites B and C. The conformation of the GlcNAc moiety of the bound (GlcNAc)$_2$ at subsite C is very similar to that of GlcNAc monomer bound at subsite C, and four hydrogen bonds are formed between the GlcNAc moiety at subsite C and the lysozyme molecule: between the NH and CO of the N-acetyl group of the sugar residue and the main chain CO of Ala 107 and NH of Asn 59, respectively, between O(6) and Trp 62 and between O(3) and Trp 63. In addition, a hydrogen bond is formed between O(3) of the GlcNAc at subsite C and O(5) of the GlcNAc at subsite B. The distance between O(6) of the GlcNAc moiety at subsite B and the carbonyl oxygen of Asp 101 is 4.7 Å, which is too large for a hydrogen bond. Such a hydrogen bond was assumed in the $(GlcNAc)_3$-hen lysozyme complex in the tetragonal form (see below). On binding of $(GlcNAc)_2$, movements of Ala 107 and Trp 62, which were very similar to those that occur upon binding of GlcNAc monomer, were observed. Therefore, hydrophobic interactions between the indole ring of Trp 62 and the sugar residue at subsite B may be important.

Blake *et al.* (1967a, b) studied the structure of $(GlcNAc)_3$-hen lysozyme complex in the tetragonal form at 2 Å resolution. $(GlcNAc)_3$ is bound to subsites A, B and C, and the interactions with the lysozyme molecule are as follows. At subsite C, the mode of binding of the sugar residue is essentially the same as that of GlcNAc bound at subsite C, and the methyl group of the N-acetyl moiety of the GlcNAc residue makes close contact with the indole ring of Trp 108. The NH and CO of this N-acetyl group are hydrogen-bonded to the main chain CO and NH groups of Ala 107 and Asn 59, respectively, and the O(6) and O(3) are hydrogen-bonded to the indole NH groups of Trp 62 and Trp 63, respectively. Asp 101 appears to be hydrogen-bonded to O(6) of the sugar residue at subsite B and to the acetamido NH of the sugar residue at subsite A (see Figs. 8-3 and 8-4). On binding of $(GlcNAc)_3$, the substrate-binding cleft is narrowed and the ring of Trp 62 moves about 0.75 Å to make non-polar contact with the pyranose ring at subsite B. In addition to these interactions, many non-polar interactions play an important role in the binding.

Crystallographic studies of the binding of GlcNAc oligomers longer than $(GlcNAc)_3$ to hen lysozyme have not been successful, since the longer oligosaccharides are hydrolyzed rapidly by lysozyme and subsites D, E and F are in contact with the neighbouring lysozyme molecules in the crystals. The mode of binding of the sugar residues to subsites D, E and F has therefore been inferred only on the basis of models of the lysozyme molecule and the saccharide. Nevertheless, the conclusions are very interesting. If a sugar residue with a normal chair conformation is placed at subsite

D, C(6) and O(6) of the sugar residue approach too closely to the main chain CO of Asp 52, and the acetamide group of the sugar residue at subsite C approaches too closely to the indole ring of Trp 108. In order to avoid such restrictions, it is necessary to modify the conformation of the sugar residue at subsite D to the half-chair conformation, in which C(1), C(2), O(5) and C(5) lie in the same

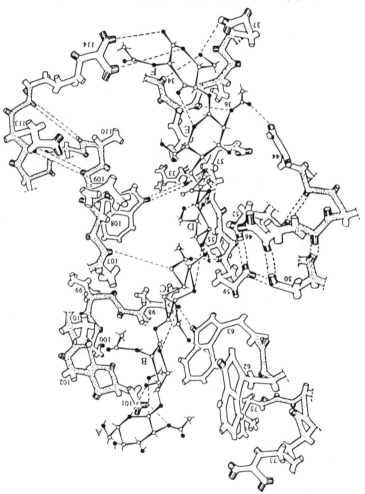

Fig. 8-3. Binding of (GlcNAc)₆ to the cleft of the hen lysozyme molecule. The polypeptide chain is shown speckled, and NH and CO are indicated by lined and full shading, respectively. (Blake, C.C.F. *et al.* (1967) *Proc. Roy. Soc., B167*, 365)

plane. With this distorted sugar residue at subsite D, hydrogen bonding between O(6) and either the main chain CO of Asp 57 or Glu 35 becomes possible. Further sugar residues can then fit at subsites E and F without difficulty. The interactions between $(GlcNAc)_6$ and hen lysozyme are shown in Figs. 8-3 and 8-4.*

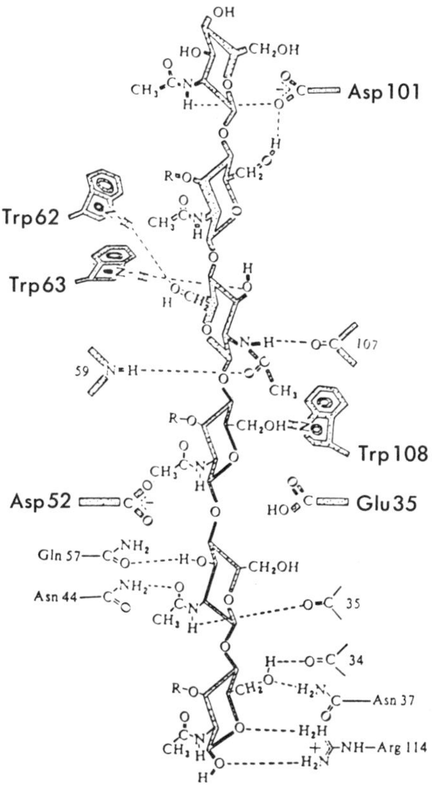

Fig. 8-4. Schematic drawing of the interactions of $(GlcNAc)_6$ with hen lysozyme. $-OR = -O-CH(CH_3)-COOH$. (modified from Dickerson and Geis, (1969) "The Structure and Action of Proteins," W. Benjamin Inc., Menlo Park, Calif.)

---

*Very recently Strynadka and James (1991) published detailed X-ray crystallographic data on the structure of MurNAc-GlcNAc-MurNAc bound to subsites B, C and D in the active-site cleft of hen lysozyme. They provide direct evidence that the MurNAc residue in subsite D is distorted from the chair conformation to a sofa conformation, thus confirming the first prediction made by Phillips 25 years ago. The hydrogen-bonding interactions between the sugar residues and protein molecule in subsites B, C and D are shown in Fig. 8-5.

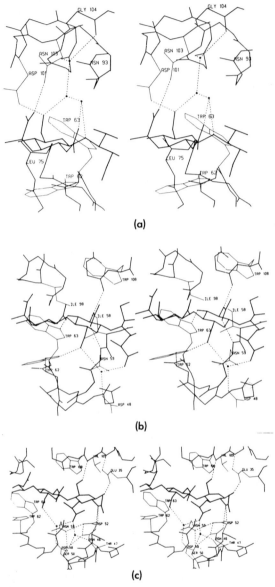

(a)

(b)

(c)

Fig. 8-5. The hydrogen bonding interactions between protein atoms and MurNAc-GlcNAc-MurNAc. (a) MurNAc residue in subsite B. (b) GlcNAc residue in subsite C. (c) MurNAc residue in subsite D. Thin lines show lysozyme side chain and thicker lines show sugar residues. Hydrogen bonds are shown as broken lines. Filled circles represent water molecules. (Strynadka, N.C.J. & James, M.N. G. (1991) *J. Mol. Biol., 220*, 401)

The mode of binding of cell-wall oligosaccharides, which contain alternating GlcNAc and MurNAc residues, to hen lysozyme has also been examined by model-building (Blake *et al.*, 1967a,b). It was suggested that MurNAc residues cannot bind to subsites A, C or E, since the MurNAc residue contains a lactyl group at position C(3) and cannot fit any of these subsites. On the other hand, a MurNAc residue can bind to subsite B, D or F, since the lactyl group can point out of the cleft. Thus, in the binding of cell-wall oligosaccharides, subsites B, D and F would be occupied by MurNAc residues and subsites A, C and E by GlcNAc residues.

In the neighbourhood of subsites D and E, the carboxyl groups of Asp 52 and Glu 35 are the most reactive groups, being disposed on either side of the β-(1→4) linkage between the residues at D and E. Asp 52 has a number of polar neighbours and appears to lie in a network of hydrogen bonds linking it with residues 46 and 59. On the other hand, Glu 35 is located in a predominantly non-polar environment (Phillips, 1967; Blake *et al.*, 1967a,b). It is known that lysozyme cleaves the β-(1→4) linkage between MurNAc and GlcNAc, but not the linkage between GlcNAc and MurNAc. On the basis of the above-mentioned mode of binding of cell-wall oligosaccharides and the fact that $(GlcNAc)_3$ occupies subsites A, B and C and forms a stable complex with lysozyme, it has been proposed that the β-(1→4) linkage between the sugar residues bound at subsites D and E is hydrolyzed by the concerted action of Glu 35 and Asp 52.

8-2. IONIZATION BEHAVIOUR OF THE CATALYTIC GROUPS

In order to understand the catalytic mechanism of an enzyme, it is essential to establish the ionization behaviour of the catalytic groups. Parsons and Raftery (1972a, b, c) first attempted to estimate the ionization constants of the catalytic groups, Asp 52 and Glu 35, of hen lysozyme. Parsons *et al.* (1969) and Parsons and Raftery (1969) succeeded in specific modification of Asp 52 using triethyloxonium fluoroborate. They measured the pH difference titration of the Asp 52-esterified lysozyme relative to the native lysozyme, and analyzed it by assuming that the ionization of ionizable groups other than Asp 52 is unaffected in the esterified lysozyme and that the ionization of Asp 52 and Glu 35 can be treated as ionization of a dibasic acid (see Eq. (4.19))

$$
\begin{array}{ccc}
 & {}_{52}E_{35} & \\
k_{1,52} \nearrow & & \nwarrow k_{1,35} \\
{}_{52^-}E_{35} & & {}_{52}E_{35^-} \\
k_{2,35} \searrow & & \swarrow k_{2,52} \\
 & {}_{52^-}E_{35^-} &
\end{array}
\tag{8.1}
$$

where E represents the lysozyme molecule, the subscripts 52 and 35 denote the protonated forms of Asp 52 and Glu 35, and 52$^-$ and 35$^-$ denote their ionized forms. $k_{1,52}$, $k_{1,35}$, $k_{2,52}$ and $k_{2,35}$ are the microscopic ionization constants.

The following equations hold for the ionization of the dibasic acid shown above (see Eqs. (4.37), (4.40) and (4.41))

$$
\left.
\begin{array}{l}
K_{52} = k_{1,52} + k_{1,35} \\[4pt]
\dfrac{1}{K_{35}} = \dfrac{1}{k_{2,35}} + \dfrac{1}{k_{2,52}}
\end{array}
\right\}
\tag{8.2}
$$

$$
\left.
\begin{array}{l}
G_1 + G_2 = k_{1,52} + k_{1,35} \\[4pt]
G_1 G_2 = k_{1,52} k_{2,35} = k_{1,35} k_{2,52}
\end{array}
\right\}
\tag{8.3}
$$

where K_{52} and K_{35} are the macroscopic ionization constants of the catalytic groups, and G_1 and G_2 are the titration constants which are the hypothetical ionization constants for the two ionizable groups of the dibasic acid, calculated as if they were an equivalent mixture of two simple monobasic acids.

If the two catalytic carboxyls interact electrostatically, then the difference titration may be analysed using the following equation

$$
\Delta\bar{h} = \frac{G_1}{(H^+) + G_1} + \frac{G_2}{(H^+) + G_2} - \frac{k_{1,35}}{(H^+) + k_{1,35}},
\tag{8.4}
$$

where $\Delta\bar{h}$ is the difference in proton release, and $k_{1,35}$ is the ionization constant of Glu 35 in Asp 52-esterified lysozyme, corresponding to the microscopic ionization constant of Glu 35 when Asp 52 has been protonated in the unmodified lysozyme. The difference titration curve was found to be best fitted using $pG_1 = 4.40 \pm 0.05$, $pG_2 = 6.10 \pm 0.05$ and $pk_{1,35} = 5.20 \pm 0.05$. The other three microconstants and two macroconstants can be calculated using Eqs. (8.2) and (8.3). Parsons and Raftery obtained 5.9 for pK_{35} and 4.5 for pK_{52} at 25°C and in 0.15 M KCl.

The CD spectra of hen, turkey and human lysozymes show a CD band at 305 nm at neutral pH values (Fig. 8-6), although their primary structures are different. This band suggests, however, that there is a tryptophyl residue common to these lysozymes. The pH difference absorption spectra of these lysozymes also show a peak at 301 nm. The fluorescence spectra under excitation at 280 nm are different from those at 305 nm. The changes in the CD band at 305 nm on addition of metal ions such as Co^{2+} and Mn^{2+}, which are known to bind at the catalytic groups, are very similar to the changes occurring concomitant with pH in this CD band (Ikeda & Hamaguchi, 1972, 1973a; Nakae et al., 1972, 1973). All of these facts suggest that the absorption band at 305 nm originates from Trp 108, which is very close to Glu 35, and this was eventually verified by detailed analysis of the pH dependence of the CD band at 305 nm, the extinction at 301.5 nm and the fluorescence at 340

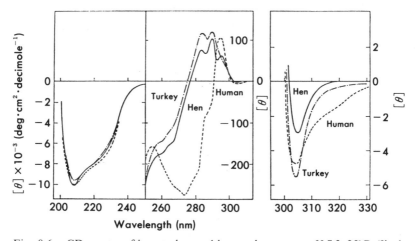

Fig. 8-6.   CD spectra of hen, turkey and human lysozymes. pH 7.2, 25°C. (Ikeda, K. & Hamaguchi, K. (1972) *J. Biochem.*, *71*, 265; Ikeda, K. *et al.* (1972) *J. Biochem.*, *71*, 371; Kuramitsu, S. *et al.* (1974) *J. Biochem.*, *76*, 671)

210

nm upon excitation at 305 nm (Kuramitsu *et al.*, 1974, 1977, 1978; Itani *et al.*, 1975).

As an example, the pH dependence curves of the extinction difference ($\Delta\varepsilon$) at 301.5 nm for intact hen lysozyme and Asp 52-esterified lysozyme are shown in Fig. 8-7. The pH dependence of $\Delta\varepsilon$ for intact lysozyme shows four transitions: below pH 2, between pH 2 and 4.5, between pH 4.5 and 7.5, and above pH 7.5. The transition below pH 2 may be due to the contribution of Asp 66 (Imoto *et al.*, 1972). Tyrosyl ionization contributes to the value of $\Delta\varepsilon$ at 301.5 nm above pH 7. While the pH dependence of $\Delta\varepsilon$ at 301.5 nm for intact lysozyme shows two transitions between pH 2 and 7.5, that for Asp 52-esterified lysozyme shows only a single transition with a midpoint at pH 5.25 in this pH region. This suggests that the ionization of Asp 52 and Glu 35 affects the extinction at 301.5 nm due to Trp 108, and that there is an electro-static interaction between Asp 52 and Glu 35.

The following scheme for a monobasic acid may be applied to the pH dependence of $\Delta\varepsilon$ at 301.5 nm of Asp 52-esterified lysozyme in the pH region of 2 to 7

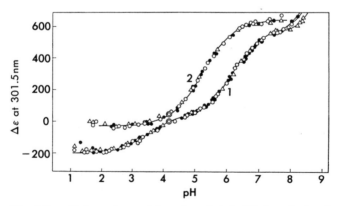

Fig. 8-7. pH dependence of the value of at $\Delta\varepsilon$ 301.5 nm for hen lysozyme (1) and Asp 52-esterified lysozyme (2) at ionic strength 0.1 and 6°C (△), 25°C (○), and 35°C (●). The reference pH is 4.18. The solid lines indicate theoretical curves. (Kuramitsu, S. *et al.* (1977) *J. Biochem., 82,* 535)

$$_eE_{35} \underset{}{\overset{K}{\rightleftharpoons}} {_eE_{35^-}} + H^+, \tag{8.5}$$

where $_eE$ represents Asp 52-esterified lysozyme, and $_eE_{35}$ and $_eE_{35^-}$ are the protonated and deprotonated forms of Glu 35, respectively, in the esterified lysozyme, and K is the ionization constant.

By assuming that the extinction of the lysozyme molecule with different ionization forms is directly proportional to the fraction of the respective lysozyme molecules, the extinction coefficient (ε) of the esterified lysozyme at a given pH is expressed by

$$\varepsilon = \varepsilon_{eEH} \frac{(_eE_{35})}{(E)_0} + \varepsilon_{eE} \frac{(_eE_{35^-})}{(E)_0} = \varepsilon_{eEH} + (\varepsilon_{eE} - \varepsilon_{eEH}) \frac{K}{(H^+) + K}, \tag{8.6}$$

where $(E)_0$ is the total concentration of Asp 52-esterified lysozyme, and ε_{eEH} and ε_{eE} are the extinction coefficients of $_eE_{35}$ and $_eE_{35^-}$, respectively. The apparent pK value of Glu 35 in the Asp 52-esterified lysozyme was determined to be 5.25 at 0.1 ionic strength and 25°C.

The pH dependence of $\Delta\varepsilon$ at 301.5 nm for intact hen lysozyme may be analysed by applying the scheme in Eq. (8.1). The molar extinction coefficient at a given pH can be expressed as follows

$$\varepsilon = \varepsilon_{HEH} \frac{(_{52}E_{35})}{(E)_0} + \varepsilon_{EH} \frac{(_{52}E_{35})}{(E)_0} + \varepsilon_{HE} \frac{(_{52}E_{35^-})}{(E)_0} + \varepsilon_E \frac{(_{52^-}E_{35^-})}{(E)_0}$$

$$= \frac{\varepsilon_{HEH} \dfrac{(H^+)^2}{k_{1,52}k_{2,35}} + \varepsilon_{EH} \dfrac{(H^+)}{k_{2,35}} + \varepsilon_{HE} \dfrac{(H^+)}{k_{2,52}} + \varepsilon_E}{\dfrac{(H^+)^2}{k_{1,52}k_{2,35}} + \dfrac{(H^+)}{k_{2,35}} + \dfrac{(H^+)}{k_{2,52}} + 1} \tag{8.7}$$

where ε_{HEH}, ε_{EH}, ε_{HE} and ε_E are the molar extinction coefficients of $_{52}E_{35}$, $_{52^-}E_{35}$, $_{52}E_{35^-}$ and $_{52^-}E_{35^-}$, respectively.

The microscopic ionization constants are linked by Eq. (8.3) and the macroscopic ionization constants of the catalytic groups, K_{52} and K_{35}, are related to the microconstants by Eq. (8.2). Using the data in Fig. 8-7, the apparent macroscopic ionization constants were determined as $pK_{35} = 6.1$ and $pK_{52} = 3.4$ at 0.1 ionic strength and 25°C.

The pH dependence curves of the ellipticity at 304 nm and the fluorescence intensity at 340 nm upon excitation at 305 nm may be analysed in the same way as done in the analysis of the extinction coefficient at 301 nm. The macroscopic ionization constants of Asp 52 and Glu 35 determined by the three methods are given in Table 8-1. This table also includes the pK values of Asp 52 and Glu 35 for turkey egg-white lysozyme and human lysozyme. The macroconstants of the catalytic groups obtained by these three methods are in good agreement with each other. Although the pK_{52} values for the three lysozymes are all the same, the pK_{35} value for human lysozyme is higher by 0.7 pH unit than that for hen or turkey lysozyme. The pK_{52} value is much lower than the normal pK value (4.0) for the β-COOH of Asp and the pK_{35} value is much greater than the normal pK value (4.5) for the γ-COOH of Glu (see Table 1-2). This seems to be consistent with the X-ray crystallographic data indicating that Asp 52 and Glu 35 are located in hydrophilic and hydrophobic environments, respectively.

The macroscopic pK value of Asp 52 in hen lysozyme determined by Parsons and Raftery (1972a,b,c) using the pH difference titration was 4.5 (see above), which is higher by about one pH unit than the pK value determined by the spectroscopic method. Fukae *et al.* (1979) pointed out that the pH difference titration data can be consistently interpreted in terms of the pK values of Asp 52 and Glu 35 shown in Table 8-1, if we assume that the pK value of another ionizable group (probably Asp 48) is perturbed on esterification of Asp 52.

By measuring the $\Delta\varepsilon$ values at 301.5 nm at various temperatures, the thermodynamic parameters for each ionization step of the

TABLE 8-1

Macroscopic pK Values of the Catalytic Groups, Asp 52 and Glu 35, for Hen, Turkey and Human Lysozymes, Determined from the Extinction Difference at 300 nm, the CD Band at 305 nm and the Fluorescence at 340 nm upon Excitation at 305 nm

Lysozyme	Difference absorption		Circular dichroism		Fluorescence	
	pK_{52}	pK_{35}	pK_{52}	pK_{35}	pK_{52}	pK_{35}
Hen	3.4	6.1	3.4	6.0	3.4	6.1
Turkey	3.4	6.1	3.4	6.0	3.4	6.1
Human	3.4	6.8	3.4	6.8	3.4	6.8

Ionic strength 0.1, 25°C.

TABLE 8-2

Thermodynamic Parameters for the Ionization of the Catalytic Groups, Asp 52 and Glu 35, of Hen Lysozyme

	pK	ΔG° (kcal/mol)	ΔH° (kcal/mol)	ΔS° (cal/deg/mol)
$k_{1,52}$	3.45	4.71	~0	−16
$k_{2,52}$	4.30	5.87	~0	−20
$k_{1,35}$	5.25	7.16	~0	−24
$k_{2,35}$	6.10	8.32	~0	−28

Ionic strength 0.1, 25°C.

TABLE 8-3

Macroscopic pK Values for the Catalytic Groups, Asp 52 and Glu 35, and Asp 101 of Hen, Turkey and Human Lysozymes and Their Complexes with $(GlcNAc)_3$, and pK Values of Glu 35 (pK'_{35}) of Asp 52-esterified Hen Lysozyme and Its Complex with $(GlcNAc)_3$

Lysozyme	pK_{52}	pK_{35}	pK_{101}	pK'_{35}
Hen	3.4	6.0	4.5	
Hen-$(GlcNAc)_3$ complex	3.4	6.5	3.4	
Turkey	3.4	6.0		
Turkey-$(GlcNAc)_3$ complex	3.4	6.5		
Human	3.4	6.8		
Human-$(GlcNAc)_3$ complex	3.4	6.7		
Asp 52-esterified hen				5.25
Asp 52-esterified hen-$(GlcNAc)_3$ complex				5.10

Ionic strength 0.1, 25°C.
Kuramitsu, S. *et al.* (1974) *J. Biochem.*, *76*, 671; Kuramitsu, S. *et al.* (1975) *J. Biochem.*, *77*, 291.

catalytic groups were determined by Kuramitsu *et al.* (1977) (Table 8-2).

In essentially the same way, the ionization constants (K_{52}^{ES} and K_{35}^{ES}) of the catalytic groups in the lysozyme-$(GlcNAc)_3$ complex can be determined by measuring the pH dependence of the ellipticity at 305 nm of lysozyme in the presence of a high concentration of the saccharide which is sufficient to saturate the lysozyme molecule. The results are summarized in Table 8-3. As described above, the macroscopic pK values, pK_{52}^{E} and pK_{35}^{E}, of saccharide-free hen lysozyme are 3.4 and 6.0, respectively. The pK_{35} value, which mainly reflects the ionization of Glu 35 when $\overline{\text{Asp } 52}$ has been ionized, shifts by 0.5 pH unit upon saccharide binding, while the pK_{52} value, which mainly reflects the ionization of Asp 52 when Glu 35 has been protonated, remains unchanged. The pK value of Glu 35 in Asp 52-esterified lysozyme is almost unaltered when complexed with $(GlcNAc)_3$. The macroscopic ionization constants of saccharide-free turkey lysozyme and of the turkey-$(GlcNAc)_3$ are the same as those for hen lysozyme. However, in the case of human lysozyme, the binding of $(GlcNAc)_3$ does not affect any of the macroconstants.

8-3.   INTERACTIONS WITH SUBSTRATE ANALOGUES

A) *Methods for Determining the Binding Constants*

The binding of substrate analogues to lysozyme has been studied extensively by various methods. In the binding of one mole of substrate (S) to one mole of enzyme (E)

$$E + S \rightleftharpoons ES, \tag{8.8}$$

the binding constant (K_{ES}) may be expressed by the following equation

$$K_{ES} = \frac{(ES)}{(E)(S)} = \frac{(ES)}{((E)_o - (ES))((S)_o - (ES))}, \tag{8.9}$$

where $(E)_o$ and $(S)_o$ are the total concentrations of the enzyme and substrate, respectively, and (E), (S) and (ES) are the equilibrium concentrations of the enzyme, substrate and enzyme-substrate complex, respectively.

In equilibrium dialysis experiments, the equilibrium concentrations of E, S and ES can be determined directly and the binding constants can be calculated relatively easily. However, when large amounts of sample are not available or the change in substrate concentration on binding to the enzyme cannot be determined precisely, other methods must be used. Furthermore, this method is relatively time-consuming, and is therefore not readily applicable to hydrolyzable substrates.

Since tryptophyl residues are present in the substrate-binding cleft of the lysozyme molecule, the binding constants can be determined by measuring changes in the ultraviolet absorption, fluorescence and CD due to tryptophyl residues on the binding of substrate. NMR is also very useful for determining the binding constants. In this case, the changes in the chemical shifts of the atomic groups of the enzyme and substrate can be utilized. When these methods are applied, the binding constants are determined by the following procedure, since the equilibrium concentrations of enzyme and substrate are unknown. Assuming that the changes in a physical parameter (y) such as the ultraviolet absorption, fluorescence intensity or ellipticity are directly proportional to the fractions of lysozyme associated with saccharide, then the observed parameter may be expressed as

$$
\left.
\begin{aligned}
\Delta y &= \Delta y_{ES} \frac{(ES)}{(E)_0} \\
\Delta y &= y - y_{ES} \\
\Delta y_{ES} &= y_{ES} - y_E
\end{aligned}
\right\} \quad (8.10)
$$

where y_E and y_{ES} represent y for the enzyme and enzyme-substrate complex, respectively.

Δy is given by

$$
\Delta y = \Delta y_{ES} \frac{K_{ES}(S)}{1 + K_{ES}(S)} \quad (8.11)
$$

or

$$
y = y_E + (y_{ES} - y_E) \frac{K_{ES}(S)}{1 + K_{ES}(S)}. \quad (8.12)
$$

(S) is related to the binding constant, K_{ES}, the total concentration of substrate, $(S)_0$, and the total concentration of enzyme, $(E)_0$, by Eq. (8.13) under the conditions, $0 < (ES) < (E)_0$ and $0 < (ES) < (S)_0$, which are satisfied in all the experiments.

$$
\begin{aligned}
(S) = \frac{1}{2} \Big\{ &(S)_0 - (E)_0 - 1/K_{ES} \\
&+ \sqrt{\{(S)_0 - (E)_0 - 1/K_{ES}\}^2 + 4(S)_0/K_{ES}} \Big\}
\end{aligned} \quad (8.13)
$$

The parameters, y_E, y_{ES} and K_{ES} are determined by fitting Eqs. (8.12) and (8.13) to the observed y at various $(S)_o$ and $(E)_o$ values using a least-squares computer program for nonlinear functions.

Equation (8.12) can be rewritten as

$$\frac{\Delta y}{(S)} = K_{ES}(\Delta y_{ES} - \Delta y_E). \qquad (8.14).$$

The standard free energy change of binding is obtained from the equation

$$\Delta G^o = -RT \ln K_{ES}. \qquad (8.15)$$

The unitary free energy change is expressed by

$$\Delta G_u = \Delta G^o - 2{,}400 \, \text{cal/mol} \qquad (8.16)$$

(see Section 1-3).

B) pH Dependence of the Binding Constants of Substrate Analogues

As described in Section 8-1, the substrate-binding site of hen lysozyme consists of six subsites, A to F. GlcNAc binds at subsite C, $(GlcNAc)_2$ binds at subsites B and C, and $(GlcNAc)_3$ binds at subsites A, B and C. In the binding site, catalytic groups (Asp 52 and Glu 35), Asp 101, Trp 62, Trp 63 and Trp 108 participate in the binding of substrate analogues.

Figure 8-8 shows the pH dependence curves for the binding constants of $(GlcNAc)_3$ to hen, turkey and human lysozymes at ionic strength 0.1 and 25°C as determined by CD measurements (Kuramitsu *et al.*, 1975a, b). Comparing first the curve for hen lysozyme with that for turkey lysozyme, in which Asp 101 in the substrate-binding site of hen lysozyme is replaced by Gly, it can be said that in the pH range below 2, the log K_{ES} values for the two lysozymes are almost the same. Between pH 2 and 5.5, the value of log K_{ES} for hen lysozyme increases, whereas that for turkey lysozyme remains unchanged. In the pH range above 5.5, a similar single transition is observed for both lysozymes. Comparison of the pH profile of the binding constant to hen lysozyme with that for turkey lysozyme suggests the participation of the ionized state of

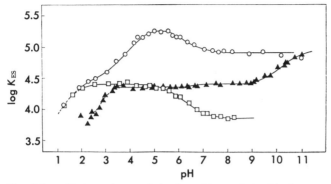

Fig. 8-8. pH dependence of the logarithms of the binding constants (K_{ES}) of (GlcNAc)$_3$ to hen lysozyme ($\circ$), turkey lysozyme ($\square$), and human lysozyme ($\blacktriangle$) at ionic strength 0.1 and 25°C. The solid lines indicate theoretical curves. (Kuramitsu, S. *et al.* (1975) *J. Biochem.*, 77, 291)

Asp 101 in the binding of (GlcNAc)$_3$ to hen lysozyme, together with a similarity in the mode of binding of this saccharide to both lysozymes except for the participation of Asp 101 in hen lysozyme. It should, therefore, be possible to interpret the pH dependence of the binding constant to turkey lysozyme in the pH range between 2 and 8 in terms of the contribution of the catalytic groups only. The binding constant of (GlcNAc)$_3$ to turkey lysozyme at a given pH value, K_{ES}, may be expressed as follows

$$\log K_{ES} = \log \frac{(_{52}E_{35}S) + (_{52^-}E_{35}S) + (_{52}E_{35^-}S) + (_{52^-}E_{35^-}S)}{((_{52}E_{35}) + (_{52^-}E_{35}) + (_{52}E_{35^-}) + (_{52^-}E_{35^-}))(S)}$$

$$= \log \frac{\dfrac{(H^+)^2}{K^{ES}_{52}K^{ES}_{35}} + \dfrac{(H^+)}{K^{ES}_{35}} + 1}{\dfrac{(H^+)^2}{K^{E}_{52}K^{E}_{35}} + \dfrac{(H^+)}{K^{E}_{35}} + 1} + \log k^E_{ES}, \qquad (8.17)$$

where K^E_{52} and K^E_{35} represent the macroscopic ionization constants of Asp 52 and Glu 35, respectively, of the saccharide-free lysozyme, and K^{ES}_{52} and K^{ES}_{35} the macroscopic ionization constants of Asp 52 and Glu 35, respectively, of the lysozyme-saccharide complex. k^E_{ES} is the binding constant to a microscopic protonation form of the turkey lysozyme molecule in which Asp 52 and Glu 35 are both ionized ($_{52^-}E_{35^-}$), and (H$^+$) is the activity of hydrogen ions.

Using the values of K^E_{52}, K^E_{35}, K^{ES}_{52} and K^{ES}_{35} (Tables 8-1 and

8-3), and assuming a value for log k_{ES}^E (3.83) which is practically the same as the binding constant at a neutral pH value, a theoretical curve can be constructed. As can be seen from Fig. 8-8, the theoretical curve (solid line) is in good agreement with the experimental data. This indicates that no ionizable groups other than the catalytic groups participate in the binding of $(GlcNAc)_3$ to turkey lysozyme in the pH range from 2 to 9.

The binding constant of $(GlcNAc)_3$ to hen lysozyme at a given pH value may be expressed in the form

$$\log K_{ES} = \log \frac{\dfrac{(H^+)^2}{K_{52}^{ES} K_{35}^{ES}} + \dfrac{(H^+)}{K_{35}^{ES}} + 1}{\dfrac{(H^+)^2}{K_{52}^{E} K_{35}^{E}} + \dfrac{(H^+)}{K_{35}^{E}} + 1} + \sum_{i=1}^{n} \log \frac{\dfrac{(H^+)}{K_i^{ES}} + 1}{\dfrac{(H^+)}{K_i^{E}} + 1} + \log k_{ES}^E, \quad (8.18)$$

where K_i^E and K_i^{ES} represent the ionization constant of an ionizable group i of lysozyme and its complex with $(GlcNAc)_3$, respectively; this group is assumed to be ionized independently of the ionization state of the catalytic groups and other groups which participate in the saccharide binding. k_{ES}^E is the binding constant of $(GlcNAc)_3$ to a microscopic protonation form of lysozyme in which all the ionizable groups participating in the binding have been ionized. As shown by X-ray crystallographic studies, Asp 101 in addition to the catalytic groups is involved in the saccharide binding. The most probable pH dependence curve can be obtained by assuming a value for the ionization constant of Asp 101 in hen lysozyme and for that in its saccharide complex, together with a value for log k_{ES}^E (4.90), and using the ionization constants of Asp 52 and Glu 35 shown in Tables 8-1 and 8-3. The values of pK_{101}^E and pK_{101}^{ES} so determined were 4.5 and 3.4, respectively.

The transition observed for hen and turkey lysozymes in the pH range below 2 (Fig. 8-8) may be due to the contribution of Asp 66 (Imoto *et al.*, 1972).

The pH profile of the binding constant of $(GlcNAc)_3$ to human lysozyme differs from those for hen and turkey lysozymes. The binding constants of $(GlcNAc)_3$ to human lysozyme remain practically unchanged between pH 9 and 3.5 and decrease below pH 3.5 (Fig. 8-8). Human lysozyme has Asp 52, Glu 35, and Asp

101 in the substrate-binding site. Nevertheless, the pH profile for human lysozyme shows neither the transition between pH 5 and 8 due to perturbation of Glu 35 in hen and turkey lysozymes, nor the transition between pH 5 and 2 due to perturbation of the ionization of Asp 101 in hen lysozyme. As shown in Tables 8-1 and 8-3, the pK_{35} value of human lysozyme does not change on binding (Glc-NAc)$_3$. Between pH 9 and 11, the binding constant of (GlcNAc)$_3$ to human lysozyme increases, whereas the binding constant to hen lysozyme is unchanged within this pH range. The theoretical curve constructed by Eq. (8.18), using the macroscopic pK values of Asp 52 and Glu 35 in the saccharide-free lysozyme and the complex with (GlcNAc)$_3$ (Tables 8-1 and 8-3), $pK_1^E = 10.5$, and $pK_1^{ES} = 10.0$, is in good agreement with the experimental data (Fig. 8-8). Trp 62 of hen lysozyme is replaced by Tyr in human lysozyme, and this tyrosyl residue lies much closer to the saccharide ring at subsite C than at B, with its ring parallel to the saccharide ring at subsite C in the human lysozyme-(GlcNAc)$_3$ complex (Banjard, 1973). The ionizable group with a pK of 10.5 which shifts to 10.0 upon binding of (GlcNAc)$_3$ may be Tyr 62.

In the lysozyme-(GlcNAc)$_3$ system at any given pH value, there are eight microscopic lysozyme species with respect to the catalytic carboxyls (Fig. 8-9). The fractions of these species are related to each other by the eight microscopic ionization constants

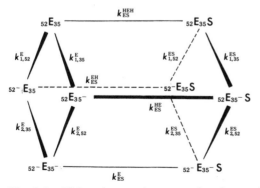

Fig. 8-9. Eight microscopic protonation forms of lysozyme in the presence of (GlcNAc)$_3$ with respect to the ionization states of Asp 52 and Glu 35.

of the catalytic groups and by the four binding constants of the saccharide to the respective microscopic protonation forms of hen, turkey and human lysozymes. The binding constants of $(GlcNAc)_3$ to each microscopic protonation form of hen lysozyme are summarized in Table 8-4.

It is interesting to note that the binding constant to the form $_{52}E_{35^-}$ of hen lysozyme is the smallest, and that those for the other species are practically the same. As will be described below, on raising the pH from 5.3 to 7.5, where the ionization process $_{52}E_{35} \rightarrow {}_{52}E_{35^-}$ occurs predominantly, the reductions in the binding constants of $(GlcNAc)_2$ and GlcNAc monomers to hen lysozyme are the same as that observed for the binding constant of $(GlcNAc)_3$. This indicates that the binding of $(GlcNAc)_2$ and GlcNAc monomers bears a similar relation to that of $(GlcNAc)_3$ to each microscopic protonation form, and suggests that a conformational change at subsite C occurs when the ionization state of the catalytic groups changes.

X-ray crystallographic studies have shown that binding of these GlcNAc oligosaccharides narrows the substrate-binding cleft, but no direct interactions occur between the sugar residue and catalytic groups (Blake et al., 1967b). This suggests that the shift of $pK_{2,52}$ or $pK_{2,35}$ observed when $(GlcNAc)_3$ binds to hen lysozyme arises, in part, because the $(GlcNAc)_3$ binding leads to a change in the microenvironment of the catalytic groups and, more significantly, because the catalytic carboxyls are drawn closer to each other and the electrostatic repulsion between them increases. This

TABLE 8-4
Binding Constants of $(GlcNAc)_3$ to Four Microscopic Protonation Forms of Lysozyme with Respect to the Catalytic Carboxyls

Lysozyme	log k_{ES}^{HEH}	log k_{ES}^{EH}	log k_{ES}^{HE}	log k_{ES}^{E}
Hen A	5.3	5.3	5.4	4.9
B	4.2	4.2	4.3	3.8
Turkey	4.3	4.3	4.4	3.8
Human	4.3	4.3	4.3	4.4

A, with Asp 101 ionized; B, with Asp 101 protonated.
Ionic strength 0.1, 25°C.
Kuramitsu, S. et al. (1975) J. Biochem., 77, 291.

appears to explain the fact that the pK_{52} value, which is the ionization constant when Glu 35 has been protonated, is almost unchanged when $(GlcNAc)_3$ binds, and that the pK value of Glu 35 in Asp 52-esterified lysozyme remains almost unaltered upon binding of $(GlcNAc)_3$. However, the binding of N^1-methylnicotinamide to Trp 62 in native hen lysozyme and in its complex with GlcNAc oligomers, the CD band at 375 nm of hen lysozyme in which Trp 62 has been selectively 2-nitrophenylsulfenylated, and the CD band at 360 nm of hen lysozyme in which Trp 62 has been selectively converted to kynurenine—these are all affected by the ionization of Glu 35 (Ikeda & Hamaguchi, 1973b, 1975; Nakae *et al.*, 1975a; Teshima *et al.*, 1980). These facts indicate that there is a relationship between the ionization of Glu 35 and the states of Trp 62 and its proximity, and that these conformational changes affect the binding of substrate analogues to hen lysozyme.

The shape of the pH profile of the binding constant of $(GlcNAc)_2$ is very similar to that of $(GlcNAc)_3$ (Nakae *et al.*, 1976), but the pK shift of Asp 101 caused by binding of the dimer is smaller than the shift for the binding of the trimer. This corresponds to the X-ray crystallographic finding that $(GlcNAc)_3$ interacts with Asp 101 at both subsites A and B, whereas $(GlcNAc)_2$ interacts with Asp 101 only at subsite B (Blake *et al.*, 1967b). The ionization

TABLE 8-5
Ionization Constants of Ionizable Groups Which Participate in the Saccharide Binding

	pK_{52}	pK_{35}	pK_{101}
Hen-$(GlcNAc)_3$	3.4	6.5	3.4
Hen-$(GlcNAc)_2$	3.4	6.5	4.0
Hen-α-GlcNAc	3.6	6.5	
Hen-β-GlcNAc	3.8	6.5	
Hen-α-Methyl-GlcNAc	3.4	6.6	
Hen-β-Methyl-GlcNAc	3.8	6.6	
Turkey-$(GlcNAc)_3$	3.4	6.5	
Turkey-α-GlcNAc	3.5	6.4	
Turkey-β-GlcNAc	3.8	6.5	
Turkey-α-Methyl-GlcNAc	3.4	6.5	
Turkey-β-Methyl-GlcNAc	3.8	6.6	
Human-$(GlcNAc)_3$	3.4	6.7	

TABLE 8-6

Binding Constants and Changes in Unitary Free Energy for Interactions of $(GlcNAc)_3$, $(GlcNAc)_2$, GlcNAc and β-Methyl-GlcNAc with Hen, Turkey and Human Lysozymes at 0.1 Ionic Strength and 25°C

Enzyme saccharide	pH 2.4–2.6		pH 5.1–5.4		pH 7.3–7.5	
	K_{ES} (M^{-1})	$-\Delta G_u$ (kcal/mol)	K_{ES} (M^{-1})	$-\Delta G_u$ (kcal/mol)	K_{ES} (M^{-1})	$-\Delta G_u$ (kcal/mol)
Hen $(GlcNAc)_3$	3.0×10^4	8.49	1.6×10^5	9.49	8.2×10^4	9.07
$(GlcNAc)_2$	2.0×10^3	6.88	3.7×10^3	7.25	2.0×10^3	6.89
GlcNAc	30	4.39	40	4.56	23	4.23
β-Me-GlcNAc	41	4.57	34	4.47	16	4.02
Turkey $(GlcNAc)_3$	2.5×10^4	8.37	2.0×10^4	8.23	7.1×10^3	7.63
$(GlcNAc)_2$	2.5×10^3	6.97	2.2×10^3	6.94	1.0×10^3	6.48
GlcNAc	33	4.45	34	4.47	17	4.07
β-Me-GlcNAc	61	4.81	31	4.41	12	3.87
Human $(GlcNAc)_3$	9.4×10^3	7.79	2.4×10^4	8.35	2.5×10^4	8.37
$(GlcNAc)_2$			6.1×10^2	6.18	8.1×10^2	6.35
GlcNAc	2.0	2.80	2.8	2.99	2.2	2.84

Ikeda, K. & Hamaguchi, K. (1975) *J. Biochem.*, 77, 1; Kuramitsu, S. *et al.* (1975) *J. Biochem.*, 77, 291; Kuramitsu, S. *et al.* (1975) *J. Biochem.*, 78, 327; Nakae, Y. *et al.* (1975) *J. Biochem.*, 78, 589; Nakae, Y. *et al.* (1976) *J. Biochem.*, 80, 435; Yang, Y. *et al.*, (1976) *J. Biochem.*, 80, 425.

TABLE 8-7

Binding Free Energy (kcal/mol) at Each Subsite Calculated Using the Binding Constants of $(GlcNAc)_3$, $(GlcNAc)_2$, GlcNAc and β-Methyl-GlcNAc to Hen, Turkey and Human Lysozymes at 0.1 Ionic Strength and 25°C

Lysozyme subsite		pH 2.4–2.6	pH 5.1–5.4	pH 7.3–7.5
Hen	A	1.6	2.2	2.2
	B	2.3–2.5	2.7–2.8	2.7–2.9
	C	4.4–4.6	4.5–4.6	4.0–4.2
Turkey	A	1.4	1.3	1.3
	B	2.2–2.5	2.4–2.5	2.4–2.6
	C	4.4–4.8	4.4–4.5	3.9–4.1
Human[a]	B		1.7	2.0
	C	2.8	3.0	3.0–3.4
	D		3.6	3.0–3.4

[a] It was assumed that $(GlcNAc)_2$ occupies subsites C and D in human lysozyme.

constants of ionizable groups which participate in the saccharide binding are summarized in Table 8-5.

The binding constants and free energies of $(GlcNAc)_3$, (Glc-

NAc)$_2$ and GlcNAc monomers to hen, turkey and human lysozymes are summarized in Table 8-6. On the basis of these binding
free energies, those of the sugar residue at subsites A, B and C can
be estimated by assuming the energies to be additive. The results are
shown in Table 8-7. The binding free energy at each subsite of the
hen and turkey lysozymes increases in the order A < B < C.

8-4. LYSOZYME-CATALYZED HYDROLYSIS

We have described the ionization constants of the catalytic
groups and ionizable groups which participate in the substrate
binding for lysozyme. One of the most important problems in
enzyme studies is to understand the catalytic reaction in terms of
the pK values of the catalytic groups and of the ionizable groups
in the active site.

Yang and Hamaguchi (1980a, b) prepared the 4-methylumbel-
liferyl-β-glycosides of (GlcNAc)$_2$ ((GlcNAc)$_2$-MeU), (GlcNAc)$_3$
((GlcNAc)$_3$-MeU) and of (GlcNAc)$_4$ ((GlcNAc)$_4$-MeU) (Fig. 8-1),
and studied the binding and hydrolysis of these saccharides by
lysozyme.

The binding constants can be determined by measuring
changes in the fluorescence at 375 nm and are shown in Table 8-8.
(GlcNAc)$_2$-MeU and (GlcNAc)$_3$-MeU bind mainly at subsites B, C
and D, and A, B, C and D, respectively, with the terminal MeU
group binding at subsite D. The binding and kinetic data showed
that (GlcNAc)$_4$-MeU binds to subsites A to E (productive binding)
and subsites A to D with the nonreducing sugar residue extending
beyond subsite A (nonproductive binding). The fraction of the
productive complex was 0.77 at pH 8.5.

The rates of the hydrolytic reactions of these saccharides can
be determined by measuring the release of 4-methylumbelliferone
fluorometrically. The pH dependence curves of k_{cat}, log K_m and
k_{cat}/K_m for the hydrolysis of (GlcNAc)$_3$-MeU by hen and turkey
lysozymes are shown in Figs. 8-10, 8-11 and 8-12, respectively. As
described above, the bond which is cleaved by lysozyme lies

TABLE 8-8

Binding Constants (K_{ES}) and Changes in Unitary Free Energy for the Interactions of Hen Lysozyme with $(GlcNAc)_2$-MeU, $(GlcNAc)_3$-MeU and $(GlcNAc)_4$-MeU

Saccharide	pH	Temperature (°C)	K_{ES} (M^{-1})	ΔG_u (kcal/mol)
$(GlcNAc)_2$-MeU	5.18	25	1.5×10^4	-8.1
$(GlcNAc)_3$-MeU	5.18	25	5.0×10^5	-10.2
$(GlcNAc)_3$-MeU	5.32	42	9.0×10^4	-9.7
$(GlcNAc)_3$-MeU	8.5	42	4.3×10^4	-9.2
$(GlcNAc)_4$-MeU	8.5	42	1.9×10^5	-10.1

Yang, Y. & Hamaguchi, K. (1980) *J. Biochem.*, *87*, 1003; Yang, Y. & Hamaguchi, K. (1980) *J. Biochem.*, *88*, 829.

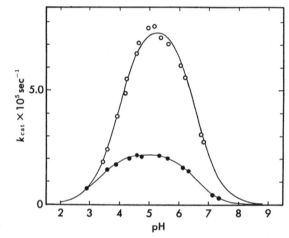

Fig. 8-10. Dependence of k_{cat} on pH for the hydrolysis of $(GlcNAc)_3$-MeU catalyzed by turkey lysozyme (○) and hen lysozyme (●). Ionic strength 0.1, 42°C. The solid lines indicate the theoretical curves. (Yang, Y. & Hamaguchi, K. (1980) *J. Biochem.*, *87*, 1003)

between subsites D and E. Therefore, $(GlcNAc)_3$-MeU should bind at subsites B, C, D and E with the terminal MeU group binding at subsite E in order to produce $(GlcNAc)_3$ and MeU hydrolytically. However, since $(GlcNAc)_3$-MeU binds mainly at subsites A, B, C and D, we must assume both non-productive and productive bind-

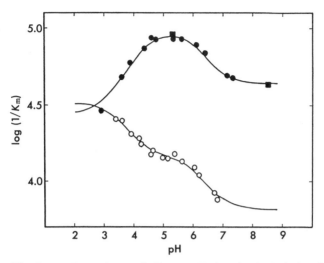

Fig. 8-11. Dependence of K_m on pH for the hydrolysis of $(GlcNAc)_3$-MeU catalyzed by turkey lysozyme (○) and hen lysozyme (●). Solid squares indicate the values of $1/K_{ES}$. Ionic strength 0.1, 42°C. The solid lines indicate the theoretical curves. (Yang, Y. & Hamaguchi, K. (1980) *J. Biochem., 87*, 1003)

ing in order to analyze the reaction kinetics. The simplest scheme may be expressed by

$$ES_{np} \underset{\longleftarrow}{\overset{K_{ES,np}}{\rightleftharpoons}} E + S \underset{\longleftarrow}{\overset{K_{ES,p}}{\rightleftharpoons}} ES_p \xrightarrow{k_2} E + P, \qquad (8.19)$$

where ES_{np} and ES_p represent non-productive ES and productive ES complexes, respectively, $K_{ES,np}$ and $K_{ES,p}$ represent the binding constants for formation of the non-productive and productive complexes, respectively, and k_2 is the rate constant.

In the scheme, productive and non-productive complexes are in rapid equilibrium with free enzyme and substrate, as indicated by the finding that the $1/K_{ES}$ value is the same as the K_m value (see Fig. 8-11), and formation of the product from the productive complex is rate-limiting. On the basis of this scheme, the apparent rate constant, k_{cat}, obtained from the Lineweaver-Burk plot may be expressed by

$$k_{cat} = \frac{K_{ES,p}}{K_{ES,p} + K_{ES,np}} \cdot k_2. \qquad (8.20)$$

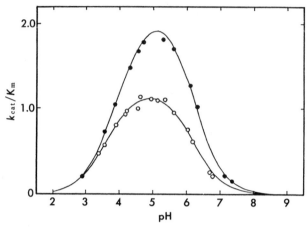

Fig. 8-12. Dependence of k_{cat}/K_m on pH for the hydrolysis of $(GlcNAc)_3$-MeU catalyzed by turkey lysozyme ($\circ$) and hen lysozyme ($\bullet$). Ionic strength 0.1, 42°C. The solid lines indicate the theoretical curves. (Yang, Y. & Hamaguchi, K. (1980) *J. Biochem., 87*, 1003)

If we assume that only the molecular species in which Asp 52 is ionized and Glu 35 is protonated is active, as suggested by the X-ray data, k_2 may be expressed by

$$k_2 = k_{max} \alpha_{52}^{ES,p} (1 - \alpha_{35}^{ES,p}), \tag{8.21}$$

where k_{max} represents the maximum rate constant, and $\alpha_{52}^{ES,p}$ and $\alpha_{35}^{ES,p}$ represent the degrees of ionization of Asp 52 and Glu 35, respectively, in the productive complex.

The binding constant, K_{ES}, of $(GlcNAc)_3$-MeU is expressed by

$$K_{ES} = K_{ES,p} + K_{ES,np}. \tag{8.22}$$

Based on these assumptions, comparison of the pH dependence curves of the kinetic constants for hen lysozyme with those for turkey lysozyme, in which Asp 101 of hen lysozyme is replaced by Gly, made it possible to determine the pK values of Asp 52, Glu 35 and Asp 101. In essentially the same way, the pK values of these ionizable groups can be obtained by analysis of the pH dependence of the kinetic constants for hydrolysis of $(GlcNAc)_4$-MeU. The results are summarized in Table 8-9.

TABLE 8-9

The pK Values of Asp 52, Glu 35 and Asp 101 of Hen and Turkey Lysozymes and Their Non-productive and Productive Complexes with (GlcNAc)$_3$-MeU and (GlcNAc)$_4$-MeU as Determined by Kinetic Measurements

		pK_{52}	pK_{35}	$pK101$
Hen-(GlcNAc)$_3$-MeU	Free enzyme	3.60	6.20	4.20
	Non-productive complex	3.95	6.55	3.30
	Productive complex	n.d.	n.d.	3.95
Hen-(GlcNAc)$_4$-MeU	Free enzyme	3.60	6.20	4.20
	Non-productive complex	3.95	6.55	3.30
	Productive complex	3.40	6.55	3.40
Turkey-(GlcNAc)$_3$-MeU	Free enzyme	3.60	6.20	—
	Non-productive complex	3.95	6.55	—

Ionic strength 0.1, 42°C.

Yang, Y. & Hamaguchi, K. (1980) *J. Biochem., 87*, 1003; Yang, Y. & Hamaguchi, K. (1980) *J. Biochem., 88*, 829.

We compare here the pK values of the catalytic groups determined using the kinetic data, and assuming that only the molecular species with ionized Asp 52 and protonated Glu 35 is active (Table 8-9), with the pK values determined independently by spectroscopic methods (Table 8-1). Although the former pK values were obtained at 42°C and the latter at 25°C, the values are comparable, since the enthalpies of ionization of Asp 52 and Glu 35 are both close to zero (Table 8-2). The macroscopic pK values of Asp 52 and Glu 35 of hen lysozyme determined by spectroscopic methods are 3.4 and 6.1, respectively, and are the same as those of turkey lysozyme. These pK values are in good agreement with those determined using the kinetic data for hydrolysis of (GlcNAc)$_3$-MeU and (GlcNAc)$_4$-MeU. This demonstrates that the catalytic mechanism of lysozyme accords with the proposal based on the X-ray data (see Section 8-1), with regard to the participation of Asp 52 and Glu 35.

The pK shift of Glu 35 of hen and turkey lysozymes on formation of the non-productive complex with (GlcNAc)$_3$ is $+0.40$ (see Table 8-3). This value is close to the pK shift observed when the non-productive complex of hen or turkey lysozyme is formed with (GlcNAc)$_3$-MeU and (GlcNAc)$_4$-MeU. On the other hand, non-productive binding of (GlcNAc)$_3$-MeU shifts the pK value of

Asp 52 by +0.35, whereas no pK shift is observed when (Glc-NAc)$_3$ binds at subsites A, B and C (Table 8-3). In the non-productive complex with (GlcNAc)$_3$-MeU, the MeU group binds at subsite D, but subsite D is not occupied by a GlcNAc residue in the complex with (GlcNAc)$_3$. Since Asp 52 is located near subsite D, this may explain the difference in the pK shift between the complex with (GlcNAc)$_3$ and the non-productive complex with (GlcNAc)$_3$-MeU.

The pK shift of Glu 35 upon formation of the productive complex with (GlcNAc)$_4$-MeU is the same as that upon formation of the non-productive complex (Table 8-8). Furthermore, this value is close to the pK shift observed when (GlcNAc)$_3$ binds to subsites A to C or GlcNAc binds to subsite C (Nakae et al., 1975b; Yang et al., 1976). This indicates that the pK shift of Glu 35 is caused by the interaction of the GlcNAc residue with subsite C and the interaction of the MeU group with subsite D in the non-productive complex, and that the interactions of the GlcNAc residue and MeU group with, respectively, subsites D and E in the productive complex are unrelated to the pK shift of Glu 35.

The pK shift of Asp 52 is observed on formation of the productive and non-productive complexes with (GlcNAc)$_4$-MeU, while the pK value of Asp 52 does not shift when (GlcNAc)$_3$ binds to subsites A to C. Thus, contrary to the pK shift of Glu 35, the pK value of Asp 52 shifts upon binding of the MeU group to subsite D or binding of the GlcNAc residue and MeU group to subsites D and E, respectively.

Books

Blundell, T.L. & Johnson, L.N. (1976) "Protein Crystallography," Academic Press, New York.

Branden, C. & Tooze, J. (1991) "Introduction to Protein Structure," Garland Publishing, Inc., New York.

Bull, H.B. (1964) "An Introduction to Physical Biochemistry," F.A. Davis Co., Philadelphia.

Creighton, T.E. (1984) "Protein Structure and Molecular Properties," W.H. Freeman & Co., New York.

Creighton, T.E. (Ed.) (1989) "Protein Structure. A Practical Approach," IRL Press at Oxford Univ. Press, Oxford.

Dickerson, R.E. & Geis, I. (1969) "The Structure and Action of Proteins," W. Benjamin Inc., Menlo Park, Calif.

Edsall, J.T. & Wyman, J. (1958) "Biophysical Chemistry," Vol. 1, Academic Press, New York.

Fasman, G.D. (Ed.) (1989) "Prediction of Protein Structure and the Principles of Protein Conformation," Plenum Press, New York.

Fersht, A. (1985) "Enzyme Structure and Mechanism," 2nd Ed., W.H. Freeman & Co., New York.

Fletterick, R.S., Schroer, T., & Matela, R.J. (1985) "Molecular Structure. Macromolecules in Three Dimensions," Blackwell Sci. Publ., Oxford.

Gierash, M.L. & King, J. (Eds.) (1990) "Protein Folding," Am. Assoc. for Adv. of Sci., Washington, D.C.

Gutfreund, H. (1972) "Enzymes: Physical Principles," John Wiley & Sons, New York.

Hiromi, K. (1979) "Kinetics of Fast Enzyme Reactions—Theory and Practice—," John Wiley & Sons, New York.

Hiromi, K., Akasaka, K., Mitsui, Y., Tonomura, B., & Murao, S. (Eds.) (1985) "Protein Proteinase Inhibitor—The Case of Streptomyces Subtilisin Inhibitor (SSI)," Elsevier Sci. Publ. B.V., Amsterdam.

Jencks, W.P. (1969) "Catalysis in Chemistry and Enzymology," McGraw-Hill Co., New York.

Oxender, D.I. & Fox, C.F. (Eds.) (1987) "Protein Engineering," Alan R. Liss, Inc., New York.

Schultz, G.E. & Schirmer, R.E. (1979) "Principles of Protein Structure," Springer-Verlag, New York.

Tanford, C. (1961) "Physical Chemistry of Macromolecules," John Wiley & Sons,

New York.

Tanford, C. (1980) "The Hydrophobic Effect: Formation of Micelles and Biological Membranes," 2nd Ed., John Wiley & Sons, New York.

Van Holde, K.E. (1971) "Physical Biochemistry," Prentice-Hall, Inc., Englewood Cliffs, N.J.

Wüthrich, K. (1986) "NMR of Proteins and Nucleic Acids," John Wiley & Sons, New York.

Reviews

CHAPTERS 2 and 3

Chothia, C. (1984) Principles that determine the structure of proteins. *Annu. Rev. Biochem., 53*, 537–572.

Chu, P.Y. & Fasman, G.D. (1978) Prediction of secondary structure of proteins from their amino acid sequence. *Adv. Enzymol., 47*, 45–148.

Go, M. (1985) Protein structure and split genes. *Adv. Biophys., 19*, 91–131.

Hol, M.G.J. (1985) The role of the α-helix dipole in protein function and structure. *Prog. Biophys. Mol. Biol., 45*, 149–195.

Johnson, W.C. Jr. (1988) Secondary structure of proteins through circular dichroism spectroscopy. *Annu. Rev. Biophys. Biophys. Chem., 17*, 145–166.

Linderstrøm-Lang, K.U. & Schellman, J.A. (1959) Protein structure and enzyme activity. "The Enzymes," 2nd Ed., Vol 1, ed. by P.D. Boyer, H. Lardy, & K. Myrbäck, Academic Press, Inc., New York, pp. 443–510.

Oppenheimer, N.J. & James, T.L. (Eds.) (1989) Nuclear magnetic resonance. Part B. Structure and Mechanism. *Methods Enzymol., 177*, 125–246.

Richards, F.M. (1977) Areas, volumes, packing and protein structure. *Annu. Rev. Biophys. Bioeng., 6*, 151–175.

Richardson, J.S. (1981) The anatomy of protein structure. *Adv. Protein Chem., 34*, 168–339.

Sutton, B.J. (1989) Immunoglobulin structure and function: the interaction between antibody and antigen. *Curr. Opinion Immunol., 2*, 106–113.

Yang, J.T., Wu, C.-S.C., & Mantinetz, H.M. (1986) Calculation of protein conformation from circular dichroism. *Methods Enzymol., 130*, 206–269.

CHAPTERS 4 and 5

Alber, T. (1989) Mutational effects on protein stability. *Annu. Rev. Biochem., 58*, 765–798.

Burley, S.K. & Petsko, G.A. (1988) Weakly polar interactions in proteins. *Adv. Protein Chem., 39*, 125–189.

Dill, K.A. (1990) Dominant forces in protein folding. *Biochemistry, 29*, 7133–7155.

Dill, K.A. & Shortle, D. (1991) Denatured states of proteins. *Annu. Rev. Biochem., 60*, 795–825.

Kauzmann, W. (1954) Denaturation of proteins and enzymes. A Symposium on the

Mechanism of Enzyme Action (Eds. W.D. McElroy & B. Glass), pp. 70–120.

Kauzmann, W. (1959) Some factors in the interpretation of protein denaturation. *Adv. Protein Chem., 14*, 1–63.

Kuwajima, K. (1989) The molten globule state as a clue for understanding the folding and cooperativity of globular-protein structure. *Proteins: Struct. Funct. Genet., 6*, 87–103.

Matthew, J.B. (1985) Electrostatic effects in proteins. *Annu. Rev. Biophys. Biophys. Chem., 14*, 387–417.

Matthew, J.B., Gurd, F.R.N., Flanagan, M.A., March, K.L., & Shire, S.J. (1985) pH-Dependent process in proteins. *CRC Crit. Rev. Biochem., 18*, 91–197.

Pace, C.N. (1975) The stability of globular proteins. *CRC Crit. Rev. Biochem., 3*, 1–43.

Pace, C.N. (1986) Determination and analysis of urea and guanidine hydrochloride denaturation curves. *Methods Enzymol., 131*, 266–280.

Pfeil, W. (1981) The problem of the stability of globular proteins. *Mol. Cell. Biochem., 40*, 3–28.

Privalov, P.L. (1979) Stability of proteins. *Adv. Protein Chem., 33*, 167–241.

Privalov, P.L. (1989) Thermodynamic problems of protein structure. *Annu. Rev. Biophys. Biophys. Chem., 18*, 47–69.

Privalov, P.L. & Gill, S.J. (1988) Stability of protein structure and hydrophobic interactions. *Adv. Protein Chem., 39*, 191–234.

Tanford, C. (1968) Protein denaturation. *Adv. Protein Chem., 23*, 121–282.

Tanford, C. (1970) Protein denaturation. *Adv. Protein Chem., 25*, 1–95.

CHAPTER 6

Huber, R. & Bode, W. (1978) Structural basis of the activation and action of trypsin. *Acc. Chem. Res., 11*, 114–122.

Hvidt, A. & Nielsen, S.O. (1966) Hydrogen exchange in proteins. *Adv. Protein Chem., 21*, 287–385.

Karplus, M. & McCammon, J.A. (1981) The internal dynamics of globular proteins. *CRC Crit. Rev. Biochem., 9*, 293–349.

Kossiakoff, A.A. (1985) The application of neutron crystallography to the study of dynamic and hydration properties of proteins. *Annu. Rev. Biochem., 54*, 1195–1227.

Wagner, G. (1983) Characterization of the distribution of internal motions in the basic pancreatic trypsin inhibitor using a large number of internal NMR probes. *Quart. Rev. Biophys., 16*, 1–57.

Woodward, C., Simon, I., & Tuchsen, E. (1982) Hydrogen exchange and the dynamic structure of proteins. *Mol. Cell. Biochem.*, *48*, 135–160.

CHAPTER 7

Anfinsen, C.B. (1973) Principles that govern the folding of protein chain. *Science*, *181*, 223–230.

Creighton, T.E. (1979) Experimental studies of protein folding and unfolding. *Prog. Biophys. Mol. Biol.*, *33*, 231–297.

Fischer, G. & Schmid, F.X. (1990) The mechanism of protein folding. Implications of *in vitro* refolding models for *de novo* protein folding and translocation in the cell. *Biochemistry*, *29*, 2205–2212.

Jaenicke, R. (1987) Folding and association of proteins. *Prog. Biophys. Mol. Biol.*, *49*, 117–237.

Jaenicke, R. (1991) Protein folding: Local structures, domains, subunits, and assemblies. *Biochemistry*, *30*, 3147–3161.

Kim, P.S. & Baldwin, R.L. (1990) Intermediates in the folding transitions of small proteins. *Annu. Rev. Biochem.*, *59*, 631–660.

Nall, B.T. (1985) Proline isomerization and protein folding. *Comments Mol. Cell. Biophys.*, *3*, 123–143.

Ptitsyn, O.B. (1987) Protein folding: Hypothesis and experiments. *J. Protein Chem.*, *6*, 273–293.

CHAPTER 8

Imoto, T., Johnson, L.N., North, A.C.T., Phillips, D.C., & Rupley, J.A. (1972) Vertebrate lysozymes. *The Enzymes* (3rd Ed.), *7*, 665–868.

References

Acharya, K.R., Stuart, D.I., Walker, N.P.C., Lewis, M., & Phillips, D.C. (1989) *J. Mol. Biol., 208*, 99-127.

Alber, T. (1989) *Annu. Rev. Biochem., 58*, 765-798.

Alber, T., Bell, J.A., Dao-pin, S., Nicholson, H., Wozniak, J.A., Cook, S., & Matthews, B.W. (1988) *Science, 239*, 631-635.

Alber, T., Dao-pin, S., Nye, J.A., Muchmore, D.C., & Matthews, B.W. (1987a) *Biochemistry, 26*, 3754-3758.

Alber, T., Dao-pin, S., Wilson, K., Wozniak, J.A., Cook, S.P., & Matthews, B.W. (1987b) *Nature, 330*, 41-46.

Anderson, D.E., Becktel, W.J., & Dahlquist, F.W. (1990) *Biochemistry, 29*, 2403-2408.

Anfinsen, C.B., Haber, E., Sela, M., & White, F.H. Jr. (1961) *Proc. Natl. Acad. Sci. U.S.A., 47*, 1309-1314.

Anson, M.L. & Mirsky, A.E. (1929) *J. Gen. Physiol., 13*, 133-143.

Ashikari, Y., Arata, Y., & Hamaguchi, K. (1985) *J. Biochem., 97*, 517-528.

Astbury, W.T. & Woods, H.J. (1930) *Nature, 126*, 913-914.

Baldwin, R.L. (1986) *Proc. Natl. Acad. Sci. U.S.A., 83*, 8069-8072.

Baldwin, R.L. (1989) *TIBS, 14*, 291-294.

Banjard, S.H. (1973) Ph. D. Thesis, Oxford Univ.

Barlow, D.J. & Thornton, J.M. (1983) *J. Mol. Biol., 168*, 867-885.

Baum, J., Dobson, C.M., Evans, P.A., & Hanley, C. (1989) *Biochemistry, 28*, 7-13.

Becktel, W.J. & Schellman, J.A. (1987) *Biopolymers, 26*, 1859-1877.

Bergman, L.W. & Kuehl, W.M. (1979) *J. Biol. Chem., 254*, 8869-8876.

Bierzyński, A., Kim, P.S., & Baldwin, R.L. (1982) *Proc. Natl. Acad. Sci. U.S.A., 79*, 2470-2474.

Blake, C.C.F., Koenig, D.F., Mair, G.A., North, A.C.T., Phillips, D.C., & Sarma, V.R. (1965) *Nature, 206*, 757-761.

Blake, C.C.F., Mair, G.A., North, A.C.T., Phillips, D.C., & Sarma, V.R. (1967a) *Proc. Roy. Soc., B167*, 365-377.

Blake, C.C.F., Johnson, L.N., Mair, G.A., North, A.C.T., Phillips, D.C., & Sarma, V.R. (1967b) *Proc. Roy. Soc., B167*, 378-388.

Blundell, T.L., Pitts, J.E., Tickle, I.J., Wood, S.P., & Wu, C.-W. (1981) *Proc. Natl. Acad. Sci. U.S.A., 78*, 4175-4179.

Brandts, J.F. & Hunt, J. (1967) *J. Am. Chem. Soc., 89*, 4826-4838.

Brandts, J.F., Halvorson, H.R., & Brennan, M. (1975) *Biochemistry, 14*, 4953-4963.

Buchner, J., Schmidt, M., Fuchs, M., Jaenicke, R., Rudolph, R., Schmid, F.X., & Kiefhaber, T. (1991) *Biochemistry, 30*, 1586-1591.

Bulleid, N.J. & Freedman, R.B. (1988) *Nature, 335*, 649-651.

Burley, S.K. & Petsko, G.A. (1985) *Science, 229*, 23-28.

Burley, S.K. & Petsko, G.A. (1988) *Adv. Protein Chem., 39*, 125-189.

Børresen, H.C. (1967) *Acta Chem. Scand., 21*, 920-936.

Campbell, R.L. & Petsko, G.A. (1987) *Biochemistry, 26*, 8579-8584.

Canfield, R.E. (1963) *J. Biol. Chem., 238*, 2698-2707.

Castellino, F.J. & Barker, R. (1968) *Biochemistry, 7*, 4135-4138.

Chance, R.E., Ellis, R.M., & Bromer, W.W. (1968) *Science, 161*, 165-167.

Chang, C.T., Wu, C.-S., & Yang, J.T. (1978) *Anal. Biochem., 91*, 13-31.

Chen, R.F. (1967) *Anal. Lett., 1*, 35-42.

Chen, B.-lu & Schellman, J.A. (1989) *Biochemistry, 28*, 685-691.

Chothia, C. (1975) *Nature, 254*, 304-308.

Chothia, C. (1984) *Annu. Rev. Biochem., 53*, 537-572.

Chou, P.Y. & Fasman, G.D. (1978) *Adv. Enzymol., 47*, 45-148.

Clore, G.M. & Gronenborn, A.M. (1991a) *J. Mol. Biol., 221*, 47-53.

Clore, G.M. & Gronenborn, A.M. (1991b) *J. Mol. Biol., 217*, 611-620.

Cooper, A. (1976) *Proc. Natl. Acad. Sci. U.S.A., 73*, 2740-2741.

Cowgill, R.W. (1967) *Biochim. Biophys. Acta, 133*, 6-18.

Creighton, T.E. (1975) *J. Mol. Biol., 95*, 167-190.

Creighton, T.E. (1983) *Biopolymers, 22*, 49-58.

Creighton, T.E. & Goldenberg, D.P. (1984) *J. Mol. Biol., 179*, 497-526.

Creighton, T.E., Hillson, D.A., & Freedman, R.B. (1990) *J. Mol. Biol., 142*, 43-62.

Cusack, S., Berthet-Colominas, C., Hartlein, M., Nassar, N., & Leberman, R. (1990) *Nature, 347*, 249-255.

Das, G., Hickey, D.R., McLendon, D., McLendon, G., & Sherman, F. (1989) *Proc. Natl. Acad. Sci. U.S.A., 86*, 496-499.

Davies, D.R., Sheriff, S., & Padlan, E.A. (1988) *J. Biol. Chem., 263*, 10541-10544.

Davies, D.R. & Padlan, E.A. (1990) *Annu. Rev. Biochem., 59*, 439-473.

De Sanctis, G., Falcioni, G., Giardina, B., Ascoli, F., & Brunori, M. (1986) *J. Mol. Biol., 188*, 73-76.

De Sanctis, G., Falcioni, G., Giardina, B., Ascoli, F., & Brunori, M. (1988) *J. Mol. Biol., 200*, 725-733.

Dill, K.A. (1990) *Biochemistry, 29*, 7133-7155.

Dill, K.A. & Shortle, D. (1991) *Annu. Rev. Biochem., 60*, 795-825.

Drexler, K.E. (1981) *Proc. Natl. Acad. Sci. U.S.A., 78*, 5275–5278.

Faber, H.R. & Matthews, B.W. (1990) *Nature, 348*, 263–266.

Fairman, R., Shoemaker, K.R., York, E.J., Stewart, J.M., & Baldwin, R.L. (1990) *Biophys. Chem., 33*, 107–119.

Fauchére, J.-L. & Pliška, V. (1983) *Eur. J. Med. Chem. Chim. Ther., 18*, 369–375.

Fischer, G. & Schmid, F.X. (1990) *Biochemistry, 29*, 2205–2212.

Fischer, G., Bang, H., & Mech, C. (1984) *Biomed. Biochim. Acta, 43*, 1101–1111.

Fischer, G., Wittmann-Liebold, B., Lang, K., Kiefhaber, T., & Schmid, F.X. (1989) *Nature, 337*, 268–270.

Flory, P.J. (1956) *J. Am. Chem. Soc., 78*, 5222–5235.

Fontana, A., Fassina, G., Vita, C., Dalzoppo, D., Zamai, M., & Zambonin, M. (1986) *Biochemistry, 25*, 1847–1851.

Frauenfelder, H., Petsko, G.A., & Tsernoglou, D. (1979) *Nature, 280*, 558–563.

Friend, S.H. & Gurd, R.R. (1979) *Biochemistry, 18*, 4612–4619.

Fukae, K., Kuramitsu, S., & Hamaguchi, K. (1979) *J. Biochem., 85*, 141–147.

Gardell, S.J., Craik, C.S., Hilvert, D., Urdea, M.S., & Rutter, W.J. (1985) *Nature, 317*, 551–555.

Gardell, S.J., Hivert, D., Barnett, J., Kaiser, E.T., & Rutter, W.J. (1987) *J. Biol. Chem., 262*, 576–582.

Garel, J.-R. & Baldwin, R.L. (1975) *J. Mol. Biol., 94*, 611–620.

Gjerde, D.T., Schmuchler, G., & Fritz, J.S. (1980) *J. Chromatogr., 187*, 35–45.

Glover, I.D., Haneef, I., Pitts, J.E., Wood, S.P., Moss, D., Tickle, I.J., & Blundell, T.L. (1983) *Biopolymers, 22*, 293–304.

Go, M. (1985) *Adv. Biophys., 19*, 91–131.

Go, M. & Nosaka, M. (1987) *Cold Spring Harbor Symp. Quant. Biol., LII*, 915–924.

Goloubinoff, P., Christeller, J.T., Gatenby, A.A., & Lorimer, G.H. (1989) *Nature, 342*, 884–889.

Goodman, E.M. & Kim, P.S. (1989) *Biochemistry, 28*, 4343–4347.

Goto, Y. & Fink, A.L. (1990) *J. Mol. Biol., 214*, 803–805.

Goto, Y. & Hamaguchi, K. (1979) *J. Biochem., 86*, 1433–1441.

Goto, Y. & Hamaguchi, K. (1981) *J. Mol. Biol., 146*, 321–340.

Goto, Y. & Hamaguchi, K. (1982a) *J. Mol. Biol., 156*, 891–910.

Goto, Y. & Hamaguchi, K. (1982b) *J. Mol. Biol., 156*, 911–926.

Goto, Y. & Hamaguchi, K. (1986) *Biochemistry, 25*, 2821–2828.

Goto, Y., Calciano, L.J., & Fink, A.L. (1990a) *Proc. Natl. Acad. Sci. U.S.A., 87*, 573–577.

Goto, Y., Ichimura, N., & Hamaguchi, K. (1988) *Biochemistry, 27*, 1670–1677.

Goto, Y., Takahashi, N., & Fink, A.L. (1990b) *Biochemistry, 29*, 3480–3488.

Gray, T.M. & Matthews, B.W. (1987) *J. Biol. Chem., 262*, 16858–16864.

Hadju, J., Acharya, K.R., Stuart, D.I., McLaughlin, P.J., Barford, D., Oikonoma-kos, N.G., Klein, H., & Johnson, L.N. (1987) *EMBO J., 6*, 539-546.

Hamaguchi, K. & Funatsu, M. (1959) *J. Biochem., 46*, 1659-1660.

Hamaguchi, K. & Geiduschek, E.P. (1962) *J. Am. Chem. Soc., 84*, 1329-1338.

Harrison, S.C. & Durbin, R. (1985) *Proc. Natl. Acad. Sci. U.S.A., 82*, 4028-4030.

Hecht, H.M., Sturtevant, J.M., & Sauer, R.T. (1986) *Proteins: Struct. Funct. Genet., 1*, 43-46.

Hol, W.G.J. (1985) *Prog. Biophys. Mol. Biol., 45*, 149-195.

Holmgren, A. & Branden, C.I. (1989) *Nature, 342*, 248-251.

Hough, E., Hansen, L.K., Birknes, B., Jynge, K., Hansen, S., Hordvik, A., Little, C., Dodson, E., & Derewenda, Z. (1989) *Nature, 338*, 357-360.

Huber, R. & Bode, W. (1978) *Acc. Chem. Rev., 11*, 114-122.

Huber, R., Kukla, D., Ruhlmann, A., Epp, O., & Fomanek, H. (1970) *Naturwissen-schaften, 57*, 389-392.

Huber, R., Kukla, D., Bode, W., Schwager, P., Bartels, K., Deisenhofer, J., & Steigemann, W. (1974) *J. Mol. Biol., 89*, 73-101.

Huber, R., Scholze, H., Paques, E.P., & Deissenhofer, J. (1980) *Hoppe-Seyler's Z. Physiol. Chem., 361*, 1389-1399.

Hughson, F.M., Wright, P.E., & Baldwin, R.L. (1990) *Science, 249*, 1544-1548.

Hvidt, A. & Nielsen, S.O. (1966) *Adv. Protein Chem., 21*, 287-385.

Ikeda, K. & Hamaguchi, K. (1972) *J. Biochem., 71*, 265-273.

Ikeda, K. & Hamaguchi, K. (1973a) *J. Biochem., 73*, 307-322.

Ikeda, K. & Hamaguchi, K. (1973b) *J. Biochem., 74*, 221-230.

Ikeda, K. & Hamaguchi, K. (1975) *J. Biochem., 77*, 1-16.

Ikeda, K., Hamaguchi, K., Miwa, S., & Nishina, T. (1972) *J. Biochem., 71*, 371-378.

Imanaka, T., Shibazaki, M., & Takagi, M. (1986) *Nature, 324*, 695-697.

Imoto, T., Johnson, L.N., North, A.C.T., Phillips, D.C., & Rupley, J.A. (1972) *The Enzymes* (3rd Ed.), *7*, 665-868.

Ishiwata, A., Kawata, Y., & Hamaguchi, K. (1991) *Biochemistry, 30*, 7766-7771.

Itani, N., Kuramitsu, S., Ikeda, K., & Hamaguchi, K. (1975) *J. Biochem., 78*, 705-711.

Jaenicke, R. (1987) *Prog. Biophys. Mol. Biol., 49*, 117-237.

Jaenicke, R. (1991) *Biochemistry, 30*, 3147-3161.

Janin, J. (1979) *Bull. Inst. Pasteur, 77*, 337-373.

Jeng, M.-F., Englander, S.W., Elöve, G.A., Wand, A.J., & Roder, H. (1990) *Biochemistry, 29*, 10433-10437.

Johnson, Jr., W.C. (1988) *Annu. Rev. Biophys. Biophys. Chem., 17*, 145-166.

Johnson, L.N. & Phillips, D.C. (1965) *Nature, 206*, 761-763.

Johnson, R.E., Adams, P., & Rupley, J.A. (1978) *Biochemistry, 17*, 1479-1484.

238

Jollès, J., Jauregui-Adell, J., Bernier, I., & Jolles, P. (1963) *Biochim. Biophys. Acta,* *78*, 668–689.

Katz, B.A. & Kossiakoff, A. (1986) *J. Biol. Chem., 261,* 15480–15485.

Kauzmann, W. (1954) A Symposium on the Mechanism of Enzyme Action (Eds. W.D. McElroy & B. Glass), pp. 70–120.

Kauzmann, W. (1959) *Adv. Protein Chem., 14,* 1–63.

Kawata, Y. & Hamaguchi, K. (1990) *Biopolymers, 30,* 389–394.

Kawata, Y. & Hamaguchi, K. (1991) *Biochemistry, 30,* 4367–4373.

Kawata, Y., Goto, Y., Hamaguchi, K., Hayashi, F., Kobayashi, Y., & Kyogoku, Y. (1988) *Biochemistry, 27,* 346–350.

Kendrew, J.C., Dickerson, R.E., Strandberg, B.E., Hart, R.G., Davies, D.R., Phillips, D.C., & Shore, V.C. (1960) *Nature, 185,* 422–427.

Kennard, C.H.L., Matsuura, Y., Tanaka, N., & Kakudo, M. (1979) *Aust. J. Chem., 32,* 911–915.

Kelly. R.F., & Richards, F.M. (1987) *Biochemistry, 26,* 6765–6774.

Kiefhaber, T., Grunert, H.-P., Harn, U., & Schmid, F.X. (1990a) *Biochemistry, 29,* 6475–6480.

Kiefhaber, T., Schmid, F.X., Renner, M., Hinz, H.-J., Hahn, U., & Quaas, R. (1990b) *Biochemistry, 29,* 8250–8257.

Kikuchi, H., Goto, Y., & Hamaguchi, K. (1986) *Biochemistry, 25,* 2009–2013.

Kim, P.S. & Baldwin, R.L. (1990) *Annu. Rev. Biochem., 59,* 631–660.

Kim, S.-H., de Vos, A., & Ogata, C. (1988) *TIBS, 13-Jan.,* 13–15.

Kline, A.D., Braun, W., & Wüthrich, K. (1986) *J. Mol. Biol., 189,* 377–382.

Klotz, I.M. (1958) *Science, 128,* 815–822.

Klotz, I.M. & Farnham, S.B. (1968) *Biochemistry, 7,* 3879–3882.

Klotz, I.M. & Franzen, J.S. (1962) *J. Am. Chem. Soc., 84,* 3461–3466.

Kossiakoff, A.A. (1982) *Nature, 296,* 713–721.

Kossiakoff, A.A. (1985) *Annu. Rev. Biochem., 54,* 1195–1227.

Kresheck, G.C. & Klotz, I.M. (1969) *Biochemistry, 8,* 8–12.

Kurachi, K., Sieker, L.C., & Jensen, L.H. (1976) *J. Mol. Biol., 101,* 11–24.

Kuramitsu, S., Ikeda, K., & Hamaguchi, K. (1975a) *J. Biochem., 77,* 291–301.

Kuramitsu, S., Ikeda, K., & Hamaguchi, K. (1975b) *J. Biochem., 78,* 327–333.

Kuramitsu, S., Ikeda, K., & Hamaguchi, K. (1977) *J. Biochem., 82,* 535–597.

Kuramitsu, S., Ikeda, K., Hamaguchi, K., Fujio, H., Amano, T., Miwa, S., & Nishina, T. (1974) *J. Biochem., 76,* 671–683.

Kuramitsu, S., Kurihara, S., Ikeda, K., & Hamaguchi, K. (1978) *J. Biochem., 83,* 159–170.

Kuroda, M., Sakiyama, F., & Narita, K. (1975) *J. Biochem., 78,* 641–651.

Kuroki, R., Taniyama, Y., Seko, C., Nakamura, H., Kikuchi, M., & Ikehara, M. (1989) *Proc. Natl. Acad. Sci. U.S.A., 86,* 6903–6907.

Kuwajima, K. (1989) *Proteins: Struct. Funct. Genet., 6,* 87–103.

Kuwajima, K., Hiraoka, Y., Ikeguchi, M., & Sugai, S. (1985) *Biochemistry, 24*, 874–881.

Kuwajima, K., Mitani, M., & Sugai, S. (1989) *J. Mol. Biol., 206*, 547–561.

Lang, K., Schmid, F.X., & Fischer, G. (1987) *Nature, 329*, 268–270.

Lapanje, S. & Tanford, C. (1967) *J. Am. Chem. Soc., 89*, 5030–5033.

Lebioda, L. & Stec, B. (1988) *Nature, 333*, 683–686.

Levitt, M. (1978) *Biochemistry, 17*, 4277–4285.

Lin, L.-N. & Brandts, J.F. (1983a) *Biochemistry, 22*, 559–563.

Lin, L.-N. & Brandts, J.F. (1983b) *Biochemistry, 22*, 564–573.

Lin, L.-N. & Brandts, J.F. (1983c) *Biochemistry, 22*, 573–580.

Lin, L.-N. & Brandts, J.F. (1983d) *Biochemistry, 22*, 4480–4485.

Lin, L.-N. & Brandts, J.F. (1984) *Biochemistry, 23*, 5713–5723.

Lin, L.-N. & Brandts, J.F. (1985) *Biochemistry, 24*, 6133–6138.

Linderstrøm-Lang, K.U. (1924) *C.R. Trav. Lab. Carlsberg, 15*, 1–29.

Linderstrøm-Lang, K. (1955) *Chem. Soc.* (Lond.) (Special publication) No. 2.

Linderstrøm-Lang, K. & Schellman, J. (1959) *The Enzymes* (2nd Ed.), *1*, 443–510.

Manavalan, P. & Johnson, Jr., W.C. (1983) *Nature, 305*, 831–832.

Marqusee, S. & Baldwin, R.L. (1987) *Proc. Natl. Acad. Sci. U.S.A., 84*, 8898–8902.

Martin, J., Langer, T., Boteva, R., Schramel, A., Horwich, A.L., & Hartl, F.-U. (1991) *Nature, 352*, 36–42.

Matsumura, M., Becktel, W.J., & Matthews, B.W. (1988a) *Nature, 334*, 406–410.

Matsumura, M., Yahanda, S., Yasumura, S., Yutani, K., & Aiba, S. (1988b) *Eur. J. Biochem., 171*, 715–720.

Matsuura, Y., Kusunoki, M., Harada, W., & Kakudo, M. (1984) *J. Biochem., 95*, 697–702.

Matthew, J.B., Gurd, F.R.N., Flanagan, M.A., March, K.L., & Shire, S.J. (1985) *CRC Crit. Rev. Biochem., 18*, 91–197.

Matthews, B.W. (1976) *Annu. Rev. Phys. Chem., 27*, 493 –523.

Matthews, B.W., Nicholson, H., & Becktel, W.J. (1987) *Proc. Natl. Acad. Sci. U.S.A., 84*, 6663–6667.

Melik-Adamyan, W.R., Barynin, V.V., Vagin, A.A., Borisov, V.V., Vainshtein, B.K., Fita, I., Murthy, M.R.N., & Rossmann, M.G. (1986) *J. Mol. Biol., 188*, 63–72.

Miller, S., Janin, J., Lesk, A.M., & Chothia, C. (1987a) *J. Mol. Biol., 196*, 641–656.

Miller, S., Lesk, A.M., Janin, J., & Chothia (1987b) *Nature, 328*, 834–836.

Mirsky, A.E. & Pauling, L. (1936) *Proc. Natl. Acad. Sci. U.S.A., 22*, 439–477.

Mitani, M., Harushima, Y., Kuwajima, K., Ikeguchi, M., & Sugai, S. (1986) *J. Biol. Chem., 261*, 8824–8829.

Moore, J.M., Peattie, D.A., Fitzgibbon, M.J., & Thomson, J.A. (1991) *Nature, 351*, 248–250.

Murphy, K.P., Privalov, P.L., & Gill, S.J. (1990) *Science, 247*, 559-561.

Nakae, Y., Ikeda, K., & Hamaguchi, K. (1973) *J. Biochem., 73*, 1249-1257.
Nakae, Y., Ikeda, K., & Hamaguchi, K. (1975a) *J. Biochem., 77*, 993-1006.
Nakae, Y., Ikeda, K., & Hamaguchi, K. (1976) *J. Biochem., 80*, 435-447.
Nakae, Y., Ikeda, K., Azuma, T., & Hamaguchi, K. (1972) *J. Biochem., 72*, 1155-1162.
Nakae, Y., Ryo, E., Kuramitsu, S., Ikeda, K., & Hamaguchi, K. (1975b) *J. Biochem., 78*, 589-597.
Nandi, P.K. & Robinson, D.R. (1972a) *J. Am. Chem. Soc., 94*, 1299-1308.
Nandi, P.K. & Robinson, D.R. (1972b) *J. Am. Chem. Soc., 94*, 1308-1315.
Neidhart, D.J., Kenyon, G.L., Gerlt, J.A., & Petsko, G.A. (1990) *Nature, 347*, 692-694.
Nettesheim, D.G., Edalji, R.P., Mollison, K.W., Greer, J., & Zuiderweg, E.R.P. (1988) *Proc. Natl. Acad. Sci. U.S.A., 85*, 5036-5040.
Nicholson, H., Becktel, W.J., & Matthews, B.W. (1988) *Nature, 336*, 651-656.
Nozaki, Y. & Tanford, C. (1963) *J. Biol. Chem., 238*, 4074-4081.
Nozaki, Y. & Tanford, C. (1965) *J. Biol. Chem., 240*, 3568-3573.
Nozaki, Y. & Tanford, C. (1967a) *J. Am. Chem. Soc., 89*, 736-742.
Nozaki, Y. & Tanford, C. (1967b) *J. Am. Chem. Soc., 89*, 742-749.
Nozaki, Y. & Tanford, C. (1967c) *J. Biol. Chem., 242*, 4731-4735.
Nozaki, Y. & Tanford, C. (1970) *J. Biol. Chem., 245*, 1648-1652.
Nozaki, Y. & Tanford, C. (1971) *J. Biol. Chem., 246*, 2211-2217.

Ohgushi, M. & Wada, A. (1983) *FEBS Lett., 164*, 21-24.
Okajima, T., Kawata, Y., & Hamaguchi, K. (1990) *Biochemistry, 29*, 9168-9175.

Pace, C.N. (1986) *Methods Enzymol., 131*, 266-280.
Pace, C.N. (1990a) *TIBS, 15-Jan.*, 14-17.
Pace, C.N. (1990b) *TIBTECH, 8*, 93-98.
Pace, C.N. & Creighton, T.E. (1986) *J. Mol. Biol., 188*, 477-486.
Pace, C.N. & Grimsley, G.R. (1988) *Biochemistry, 27*, 3242-3246.
Pace, C.N. & Laurents, D.V. (1989) *Biochemistry, 28*, 2520-2525.
Pace, C.N., Laurents, D.V., & Thomson, J.A. (1990) *Biochemistry, 29*, 2564-2572.
Pace, C.N., Shirley, B.A., & Thompson, J.A. (1989) "Protein Structure. A Practical Approach," ed. by T.E. Creighton, IRL Press at Oxford Univ. Press, Oxford, pp. 311-330.
Pace, C.N., Grimsley, G.R., Thomson, J.A., & Barnett, B.J. (1988) *J. Biol. Chem., 263*, 11820-11825.
Pantoliano, M.W., Ladner, R.C., Bryan, P.N., Rollence, M.L., Wood, J.F., & Poulos, T.L. (1987) *Biochemistry, 26*, 2077-2082.
Pantoliano, M.W., Whitelow, M., Wood, J.F., Rollence, M.L., Finzel, B.C.,

Gilliland, G.L., Poulos, T.L., & Bryan, P.N. (1988) *Biochemistry, 27*, 8311–8317.

Papiz, M.Z., Sawyer, L., Eliopoulos, E.E., North, A.C.T., Findley, J.B.C., Sivaprasadarao, R., Jones, T.A., Newcomer, M.E., & Kraulis, P.J. (1986) *Nature, 324*, 383–385.

Parsons, S.M. & Raftery, M.A. (1969) *Biochemistry, 8*, 4199–4205.

Parsons, S.M. & Raftery, M.A. (1972a) *Biochemistry, 11*, 1623–1629.

Parsons, S.M. & Raftery, M.A. (1972b) *Biochemistry, 11*, 1630–1633.

Parsons, S.M. & Raftery, M.A. (1972c) *Biochemistry, 11*, 1633–1638.

Parsons, S.M., Jao, L., Dahlquist, F.W., Borders, Jr., C.L., Groff, T., Racs, J., & Raftery, M.A. (1969) *Biochemistry, 8*, 700–712.

Pauling, L. & Corey, R.B. (1951) *Proc. Natl. Acad. Sci. U.S.A., 37*, 251–256.

Pauling, L., Corey, R., & Branson, H. (1951) *Proc. Natl. Acad. Sci. U.S.A., 37*, 205–211.

Perry, L.J. & Wetzel, R. (1986) *Biochemistry, 25*, 733–739.

Perutz, M.F., Kendrew, J.C., & Watson, H.C. (1965) *J. Mol. Biol., 13*, 669–678.

Pflugrath, J., Wiegand, G., & Huber, R. (1986) *J. Mol. Biol., 189*, 383–386.

Phillips, D.C. (1966) *Sci. Am., 215*, 78–90.

Phillips, D.C. (1967) *Proc. Natl. Acad. Sci. U.S.A., 57*, 484–495.

Phillips, D.C., Rivers, P.S., Sternberg, M.J.E., Thornton, L.M., & Wilson, I.A. (1977) *Biochem. Soc. Trans., 5*, 642–647.

Pilton, Jr., R.F., Aewan, J.C., & Petsko, G.A. (1992) *Biochemistry, 31*, 2468–2481.

Privalov, P.L. (1989) *Annu. Rev. Biophys. Biophys. Chem., 18*, 47–69.

Privalov, P.L. & Gill, S.J. (1988) *Adv. Protein Chem., 39*, 191–234.

Privalov, P.L. & Gill, S.J. (1989) *Pure Appl. Chem., 61*, 1097–1104.

Provencher, S.W. & Glöckner, J. (1981) *Biochemistry, 20*, 33–37.

Ptitsyn, O.B. (1987) *J. Protein Chem., 6*, 272–293.

Ramachandran, G.N. (1967) *"Treatise on Collagen,"* Vol. 1, ed. by G.N. Ramachandran, Academic Press, New York, p. 127.

Ramachandran, G.N. & Sasisekharan, V. (1968) *Adv. Protein Chem., 23*, 283–437.

Ramdas, L. & Nall, B.T. (1986) *Biochemistry, 25,* 6952–6964.

Richards, F.M. (1958) *Proc. Natl. Acad. Sci. U.S.A., 44*, 162–166.

Richards, F.M. (1974) *J. Mol. Biol., 82*, 1–14.

Richards, F.M. (1977) *Annu. Rev. Biophys. Bioeng., 6*, 151–175.

Richardson, J.S. (1981) *Adv. Protein Chem., 34*, 168–339.

Roder, H., Elöve, G.A., & Englander, S.W. (1988) *Nature, 335*, 700–704.

Ropson, I.J., Gordon, J.I., & Frieden, C. (1990) *Biochemistry, 29*, 8591–8599.

Roseman, M.A. (1988) *J. Mol. Biol., 201*, 621–623.

Roxby, R. & Tanford, C. (1971) *Biochemistry, 10*, 3348–3352.

Ryu, S.-E., Kwong, P.D., Truneh, A., Porter, T.G., Arthos, J., Rosenberg, M., Dai, X., Xuong, N.-H., Axel, R., Sweet, R.W., & Hendrickson, W.A. (1990) *Nature,*

348, 419-426.

Sacchettini, J.C., Meninger, T.A., Lowe, J.B., Gordon, J.I., & Banaszak, L.J. (1988) *J. Biol. Chem., 263*, 5815-5819.

Sáli, D., Bycroft, M., & Fersht, A.R. (1988) *Nature, 335*, 740-743.

Sanger, F. & Tuppy, H. (1951a) *Biochem. J., 43*, 463-481.

Sanger, F. & Tuppy, H. (1951b) *Biochem. J., 43*, 481-490.

Sauer, R.T., Jordan, S., & Palo, C.O. (1990) *Adv. Protein Chem., 40*, 1-61.

Schellman, J.A. (1975) *Biopolymers, 14*, 999-1018.

Schellman, J.A. (1978) *Biopolymers, 17*, 1305-1322.

Schiffer, M., Girling, R.L., Ely, K.R., & Edmundson, A.B. (1973) *Biochemistry, 12*, 4620-4631.

Shimizu, A., Honzawa, M., Yamamura, Y., & Arata, Y. (1980) *Biochemistry, 19*, 2784-2790.

Shoemaker, K.R., Kim, P.S., York, E.J., Stewart, J.M., & Baldwin, R.L. (1987) *Nature, 326*, 563-567.

Shoemaker, K.R., Fairman, R., Schultz, D.A., Robertson, A.D., York, E.J., Stewart, J.M., & Baldwin, R.L. (1990) *Biopolymers, 29*, 1-11.

Shoemaker, K.R., Kim, P.S., Brems, D.N., Marqusee, E.J., York, E.L., Chaiken, I. M., Stewart, J.M., & Baldwin, R.L. (1985) *Proc. Natl. Acad. Sci. U.S.A., 82*, 2349-2353.

Silverton, E.W., Navia, M.A., & Davies, D.R. (1977) *Proc. Natl. Acad. Sci. U.S.A., 74*, 5140-5144.

Sprang, S.R., Acharya, K.R., Goldsmith, E.J., Stuart, D.I., Varill, K., Fletterick, R.J., Madsen, N.B., & Johnson, L.N. (1988) *Nature, 336*, 215-221.

Sternberg, M.J.E. & Thornton, J.M. (1976) *J. Mol. Biol., 105*, 367-382.

Sternberg, M.J.E. & Thornton, J.M. (1978) *Nature, 271*, 15-20.

Strynadka, N.C.J. & James, M.N.G. (1991) *J. Mol. Biol., 220*, 401-424.

Sturtevant, J.M. (1977) *Proc. Natl. Acad. Sci. U.S.A., 74*, 2236-2240.

Sutton, B.J. (1989) *Curr. Opinion Immunol., 2*, 106-113.

Suzuki, A., Tsunogae, Y., Tanaka, I., Yamane, T., Ashida, T., Norioka, S., Hara, S., & Ikenaka, T. (1987) *J. Biochem., 101*, 267-274.

Tainer, J.A., Getzoff, E.D., Paterson, Y., Olson, A.J., & Lerner, R.A. (1985) *Annu. Rev. Immunol., 3*, 501-535.

Takagi, T. & Isemura, T. (1966) *Biochim. Biophys. Acta, 130*, 233-240.

Takahashi, N., Hayano, T., & Suzuki, M. (1989) *Nature, 337*, 473-475.

Tanaka, S., Kawata, Y., Wada, K., & Hamaguchi, K. (1989) *Biochemistry, 28*, 7188-7193.

Tanford, C. (1962) *J. Am. Chem. Soc., 84*, 4240-4247.

Tanford, C. (1968) *Adv. Protein Chem., 23*, 121-282.

Tanford, C. (1970) *Adv. Protein Chem., 25*, 1-95.

Tanford, C. & Kirkwood, J.G. (1957) *J. Am. Chem. Soc., 79*, 5333–5339.

Tanford, C. & Roxby, R. (1972) *Biochemistry, 11*, 2192–2198.

Tanford, C., Kawahara, K., & Lapanje, S. (1967) *J. Am. Chem. Soc., 89*, 729–736.

Teale, F.W. & Weber, G. (1957) *Biochem. J., 65*, 476–482.

Teeter, M.M. (1984) *Proc. Natl. Acad. Sci. U.S.A., 81*, 6014–6018.

Teeter, M.M. (1989) "Protein Folding," ed. by L.M. Gierash & J. King, Am. Assoc. Adv. Sci., Washington, D.C., pp. 44–54.

Teshima, K., Kuramitsu, S., Hamaguchi, K., Sakiyama, F., Mizuno, K., & Yamasaki, N. (1980) *J. Biochem., 87*, 1015–1027.

Thornton, J.M. (1981) *J. Mol. Biol., 151*, 261–287.

Timasheff, S.N. & Arakawa, T. (1989) "Protein Structure. A Practical Approach," ed. by T.E. Creighton, IRL Press at Oxford Univ. Press, Oxford, pp. 331–345.

Tonan, K., Kawata, Y., & Hamaguchi, K. (1990) *Biochemistry, 29*, 4424–4429.

Udgaonkar, J.B. & Baldwin, R.L. (1988) *Nature, 335*, 694–699.

von Hippel, P.H. & Schleich, T. (1969) *Acc. Chem. Res., 2*, 257–265.

Wada, A. & Ohgushi, M. (1983) *FEBS Lett., 164*, 21–24.

Wang, J., Yan, Y., Garrett, T.P.J., Liu, J., Rodgers, D.W., Garlick, R.L., Tarr, G.E., Husain, Y., Reinherz, E.L., & Harrison, S.C. (1990) *Nature, 348*, 411–418.

Washabaugh, M.W. & Collins, K.D. (1986) *J. Biol. Chem., 261*, 12477–12485.

Weaver, L.H., Grutter, M.G., Remington, S.J., Gray, T.M., Isaacs, N.W., & Matthews, B.W. (1985) *J. Mol. Evol., 21*, 97–111.

Weissman, J.S. & Kim, P.S. (1991) *Science, 253*, 1386–1393.

White, F.H. Jr. (1976) *Biochemistry, 15*, 2906–2912.

Wlodawer, A. & Sjolin, L. (1982) *Proc. Natl. Acad. Sci. U.S.A., 79*, 1418–1422.

Wood, S.P., Pitts, J.E., Blundell, T.L., Tickle, I.J., & Jenkins, J.A. (1977) *Eur. J. Biochem., 78*, 119–126.

Wright, P.E., Dyson, H.J., & Lerner, R.A. (1988) *Biochemistry, 27*, 7167–7175.

Wu, H. (1931) *Chinese J. Physiol., 5*, 321–344.

Wüthrich, K. & Wagner, G. (1978) *TIBS-Oct.*, 227–230.

Yamasaki, N., Tsujita, T., Eto, T., Masuda, S., Mizuno, K., & Sakiyama, F. (1979) *J. Biochem., 86*, 1291–1300.

Yang, Y. & Hamaguchi, K. (1980a) *J. Biochem., 87*, 1003–1014.

Yang, Y. & Hamaguchi, K. (1980b) *J. Biochem., 88*, 829–836.

Yang, Y., Kuramitsu, S., Nakae, Y., Ikeda, K., & Hamaguchi, K. (1976) *J. Biochem., 80*, 425–434.

Yutani, K., Ogasahara, K., Tsujita, T., & Sugino, Y. (1987) *Proc. Natl. Acad. Sci. U.S.A., 84*, 4441–4444.

Zimm, B.H. & Bragg, J.K. (1959) *J. Chem. Phys., 31*, 526-535.

Index